T0342369

ORGANIC
LIGHT-EMITTING
TRANSISTORS

ORGANIC LIGHT-EMITTING TRANSISTORS

Towards the Next Generation Display Technology

MICHELE MUCCINI

STEFANO TOFFANIN

Co-Published by John Wiley & Sons, Inc., Hoboken, New Jersey; and ScienceWise Publishing
Published simultaneously in Canada

For general information on our other products and services or for technical support, please contact our Customer Care Department within the United States at (800) 762-2974, outside the United States at (317) 572-3993 or fax (317) 572-4002.

Wiley also publishes its books in a variety of electronic formats. Some content that appears in print may not be available in electronic formats. For more information about Wiley products, visit our web site at www.wiley.com.

Library of Congress Cataloging-in-Publication Data

Names: Muccini, Michele, author. | Toffanin, Stefano, author.
Title: Organic light-emitting transistors : towards the next generation
display technology / Michele Muccini, Stefano Toffanin.
Description: Hoboken, New Jersey : John Wiley & Sons, 2016. | Includes
bibliographical references and index.
Identifiers: LCCN 2015042770 | ISBN 9781118100073 (cloth)
Subjects: LCSH: Electroluminescent devices–Materials. | Transistors. | Light
emitting diodes. | Organic semiconductors. | Information display
systems–Materials.
Classification: LCC TK7871.68 .M83 2016 | DDC 621.3815/28–dc23
LC record available at http://lccn.loc.gov/2015042770

Cover image courtesy of STILLFX/Getty

Set in 11/13pt Times by SPi Global, Pondicherry, India

Printed in the United States of America

10   9   8   7   6   5   4   3   2   1

CONTENTS

1

INTRODUCTION

This book is focused on organic light-emitting transistors and on their characteristics, which make them a potentially disruptive technology in a variety of application fields, including display and sensing. The distinguishing feature of this class of devices is the use of a planar field-effect architecture to combine in a single-structure electrical switching, electroluminescence generation, and photon management in organic materials.

Organic semiconductors are carbon-rich compounds with a structure tailored to optimize functions, such as charge mobility or luminescent properties. A distinguishing factor resides in the multiple functionalities that organic materials can sustain contemporarily when properly tailored in their chemical structure. This may allow the fabrication of multifunctional organic devices using extremely simple device structures and, in principle, a single active material. Indeed, in a molecular solid in which the constituting units are molecules held together by weak van der Waals forces, the optical properties are dominated by excitons, which are molecular excited states that are mobile within the solid. Excitons can hop from molecule to molecule or, in the case of polymers, from chain to chain as well as along the polymer backbone, until it recombines generating light in a radiative process. Similarly, charge carrier (electron or hole) transport can occur via hopping between molecular sites or from chain to chain. In this case, the carrier mobilities are quite low compared with inorganic semiconductors, whose room temperature values

Organic Light-Emitting Transistors: Towards the Next Generation Display Technology, First Edition. Michele Muccini and Stefano Toffanin.
© 2016 John Wiley & Sons, Inc. Published 2016 by John Wiley & Sons, Inc.

typically range from 100 to 10^4 cm^2/Vs. In contrast, in highly ordered molecular materials where charges hop between closely spaced molecules forming a crystalline stack, mobilities of less than 1 cm^2/Vs have been observed at room temperature. This is an approximate upper bound, with the mobility ultimately limited by thermal motion between neighboring molecules. Low mobility leads to low electrical conductivity and also results in a very low charge carrier velocity, which one has to consider as an intrinsic factor when evaluating the practical applications of organic semiconducting materials.

The weak van der Waals forces typical of molecular solids decrease as $1/R^6$, where R is the intermolecular spacing. This is in contrast to inorganic semiconductors that are covalently bonded, whose strength falls off as $1/R^2$. Hence, organic electronic materials are soft and flexible, whereas inorganic semiconductors are hard, brittle, and relatively robust when exposed to adverse environmental agents, such as moisture, corrosive reagents, and plasmas, commonly used in device fabrication. The apparent fragility of organic materials has also opened the door to a suite of innovative fabrication methods that are simpler to implement on a large scale than has been thought possible in the world of inorganic semiconductors.

The most appealing property of organic materials for electronic and photonic applications is that they can be deposited on virtually any substrates, including silicon backplanes and low-cost ones such as plastic, metal foils, and glass. Organic materials are compatible with low-cost fabrication methods that can be implemented on a large scale, such as vacuum sublimations and solution-based processes. This fundamental advantage and the low amount of material used in thin-film devices position them favorably to fill the application markets where cost is a key factor and the requirements on performances do not impose the use of high-performing inorganic semiconductors.

Organic electronics are beginning to make significant inroads into the commercial world, and if the field continues to progress at its current pace, electronics based on organic thin-film materials will soon become a mainstream of our technological existence. Already products based on active thin-film organic devices are in the marketplace, most notably the displays of several mobile electronic appliances. Yet, to unravel the greater promise of this technology with an entirely new generation of ultralow-cost, lightweight, and even flexible electronic devices, new and alternative solutions must be identified to overcome the limitations currently faced with the existing device architectures.

Indeed, the vertical-type structure of organic light-emitting diodes (OLEDs) is very well known and has been extremely successful for developing low-voltage-driven light-emitting devices, eventually fabricated on large-area flexible substrates. However, since OLED is a current-driven device, its application, for example, in display technology, requires high-quality TFT backplanes such as those based on LTPS—Low-Temperature Polysilicon—which increase, on the one hand, the production costs and, on the other hand, hinder the development of a truly flexible platform. On the contrary, OLET is a voltage-driven device that can be switched on and off exclusively by applying a potential, with no constrains on the current density of the switching device. This has the profound implication that lower quality TFT backplanes can be used to drive OLET frontplanes in a radically new approach toward low-cost and flexible display technology. In addition, the combination of electrical switching and light generation in a single device structure simplifies the driving circuit of the display, and therefore the manufacturing processing, ultimately leading to decreased production costs. It is also worth mentioning that OLETs offer an ideal structure for improving the lifetime and efficiency of organic light-emitting materials due to the different driving conditions with respect to standard OLED architectures and to optimized charge carrier balances.

This book aims at providing the scientific fundamentals and the key technological figures of organic light-emitting transistors (OLETs) by putting them in the context of organic electronics and benchmarking their characteristics with respect to OLEDs for applications in display and sensing technology.

In chapter 2, the OLED device features and its state-of-the-art performances are reviewed and the display technology applications are discussed. A comparative analysis of the OLED with respect to the OLET is provided to highlight the fundamental differences in terms of device architecture and working principles.

In chapter 3, the basic optoelectronic characteristics of OLETs are reported and the different structures of the active layer are correlated to the device properties.

In chapter 4, the constituting building blocks of the OLET device are discussed and their role in determining the ultimate device performance is highlighted.

In chapter 5, the charge transport and photophysical properties of OLET are analyzed, with particular emphasis on the excitonic properties and the spatial emitting characteristics.

In chapter 6, the photonic properties of OLETs are presented, focusing on the external quantum efficiency, the brightness, and the light outcoupling and emission directionality and reviewing the opportunity offered by the OLET structure for the long-searched organic injection lasing.

In chapter 7, the key applications of OLETs are discussed, driving the attention to the potential impact on display technology and sensing.

2

ORGANIC LIGHT-EMITTING DIODES

When considering devices for achieving efficient and bright electroluminescence from organic materials, it is mandatory to start any analysis from organic light-emitting diodes (OLEDs), which are by far the most advanced and developed organic electroluminescent devices. OLEDs are successfully tackling the mobile display market and are gaining momentum for general lighting applications. In this chapter, the characteristics of OLEDs in terms of device structure and working principle are outlined and the main applications of the OLED technology reviewed. By directly comparing the vertical diode architecture with the planar transistor structure, it will be clear that organic light-emitting transistors have the potential to enhance the optoelectronic performances of the photonic components, while preserving the simplicity of the system architecture at potentially lower production costs. Indeed, the combination of electronic, optoelectronic, and photonic functionalities in a single device structure has the potential to pave the way toward a novel technological platform with high integration capability and simplified manufacturing process.

2.1 OLED DEVICE STRUCTURE AND WORKING PRINCIPLES

OLED displays possess a number of advantages over conventional display devices, such as high brightness and contrast, high luminous efficiency, fast response time, wide viewing angle, low power consumption, and lightweight.

Organic Light-Emitting Transistors: Towards the Next Generation Display Technology,
First Edition. Michele Muccini and Stefano Toffanin.
© 2016 John Wiley & Sons, Inc. Published 2016 by John Wiley & Sons, Inc.

Although manufacturing costs are an issue, OLED displays can be fabricated on large-area substrates (including flexible substrates) and offer a virtually unlimited choice of colors. The technological promise of these unique characteristics puts OLEDs at the forefront of research efforts by a number of government agencies, industries, and universities. Major industrial electronics players and a number of newcomers have invested heavily in OLED research and development. As a result, a stream of new OLED products has reached the marketplace and a number of large-scale manufacturing facilities have been built or are under construction. Although the field is expected to continue growing at a rapid pace, major challenges still remain, especially the lack of highly efficient and stable organic light-emitting materials, the critical operational lifetime of the blue color, and technical hurdles in large-scale manufacturing yields of the OLED displays.

In general, light-emitting diodes (LEDs) are optoelectronic devices, which generate light when they are electrically biased in the forward direction. The early commercial LED devices, in the 1960s, were based on inorganic semiconductors such as gallium arsenide phosphide (GaAsP) as an emitter, and their efficiencies were very low. After 40 years of development, the efficiencies of inorganic LEDs have been significantly improved, and they are used in a wide range of applications such as telecommunications, indicator lights, and solid-state lightening. For flat-panel displays, the applications of LEDs have been limited to billboard displays where individual LEDs are mounted on the display boards.

Once organic thin films (either small molecules or polymers) are implemented in the diode active layer, the device is named OLED. Before the realization of the first OLED, organic-based devices could be operated only in electroluminescence mode. The first organic electroluminescence device was demonstrated in the 1950s, and very high operating voltages were required. These devices were made of anthracene single crystals doped with tetracene, which were inserted between two metal electrodes. Very high driving voltages were required and the efficiencies were very low. In the 1980s, a technological breakthrough was achieved by lowering the turn-on voltage in OLEDs. Indeed, OLEDs based on multilayer active region were fabricated consisting of a transparent anode, a hole-transporting layer, an electron-emitting layer, and a cathode. During the operation, electrons and holes are injected from the cathode and the anode, respectively, which then recombine radiatively generating light. The operation of OLEDs is similar to that of LEDs.

OLEDs are ultrathin, large-area light sources made of thin-film organic semiconductors sandwiched between two electrodes. State-of-the-art small-molecule-based OLEDs consist of various layers—each layer having a distinct functionality. These films are prepared by thermal evaporation in high

(a)　　　　　　　　　　　　　　　(b)

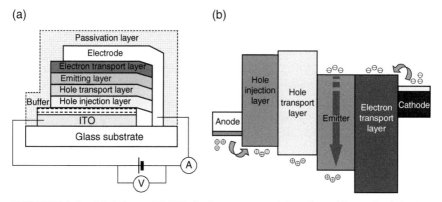

FIGURE 2.1 Multilayer OLED device structure (a) and working principle (b).

vacuum or organic vapor-phase deposition [1–3]. In contrast, polymer OLEDs are typically processed by spin-on or spray-coating techniques [4,5], where the solvent is removed by annealing steps. Polymer OLEDs are limited in their complexity owing to the fact that the solvents used often harm the underlying layers. In order to improve the general complexity of wet-processed devices, significant effort is spent on improving polymer processing.

The general architecture of an OLED, in the case of conventional bottom-emitting device, comprises a transparent electrode on top of a glass substrate (anode), followed by one or more layers of organic materials and capped with a highly reflective metal electrode (cathode). By altering the optical properties of the electrodes, top-emitting [6–8] and transparent [9] OLEDs can be fabricated. The schematic representation of a device structure and the energy level diagram of a typical multilayer OLED is reported in Figure 2.1. Firstly, efficient hole injection from the anode and efficient electron injection from the cathode are mandatory for obtaining high-efficiency devices. In inorganic semiconductors, carrier injection is achieved by heavily doping the semiconductors (n- or p-type) at the contacts to allow tunneling of the carriers through the barriers. In organic semiconductors, the optimization of injection process is obtained by matching the work-function level of the anode metal with the highest occupied molecular orbital (HOMO) of the organic semiconductor for hole injection and the work-function level of the cathode metal with the lowest unoccupied molecular orbital (LUMO) of the organic semiconductor for electron injection. The most commonly used metals and conductive oxides present work-function levels that are well aligned with the HOMO levels of the organic materials, while highly reactive low work-function metals are generally required for electron injection electrodes. To facilitate carrier injection upon the application of the external electric field, carrier injection layers with proper energy alignment with injection electrodes are necessary. Specifically, an electron injection layer (EIL) with the LUMO level matching the work-function level of the cathode is needed, while a hole injection layer (HIL) with its HOMO level matching the work-function level of the (transparent) anode is needed. To transport the injected carriers from the injection layer to the emitting layer, electron-transport layer and hole-transport layer are necessary (ETL and HTL, respectively). The migration of charge carriers (or polarons as the charge carriers are referred to when placed into a highly polarizable medium such as organic materials) occurs by means of a so-called *charge-hopping mechanism* [10] through the electron- and hole-transport materials. Ideally, the electron-transport layer should have high electron bulk mobility and the HTL should have high hole bulk mobility. In addition, these transport layers have a large energy gap in order to provide an energetically favored path for one type of charge carrier, while acting as a blocking layer for the other charge carrier. The energy level diagram of the overall system has to be designed such that the HTL presents a wide energy gap and the HOMO level matches the HOMO level of the HIL. In such configurations (Figure 2.1b), the LUMO level of the HTL is higher than the LUMO level of the ETL with the consequent formation of an energy barrier for the transport of the electrons.

The energetic constraint inherently related to the heterostructure approach is functional to the efficient light formation into the device. Indeed, the charges are favored to gather in the emission layer (EML) and recombine, an exciton is formed, and depending upon the nature of the emission materials and according to appropriate selection rules, singlet fluorescence or triplet phosphorescence is emitted. Although the structure of a typical OLED may contain many layers, not all of these layers are necessarily present in all OLED architectures. As it can be easily understood, much of the current research on OLEDs focuses on the development of the simplest possible and most easily processed architecture that can deliver the optimal combination of device properties.

Let us consider in more detail the specific characteristics required for the organic materials comprising the most important functional layers.

The HTL materials are very common in small-molecule-based OLED devices but are less common in polymer-based devices because conjugated polymers are usually good hole conductors themselves. They serve to provide a hole-conductive (via charge hopping) pathway for positive charge carriers to migrate from the anode into the EML. On the basis of this requirement, hole-transport materials are usually easily oxidized and are fairly stable in the one-electron oxidized form. This feature is related to the shallow HOMO energy level in the solid state, which is preferably isoenergetic with the anode/HIL workfunction and lower in energy than the HOMO energy level of the EML. This latter property improves the chances of charge flow into the EML with minimal charge trapping. As the main function of the HTL is to conduct the positive charge carrier holes, hole traps (higher energy HOMO materials) should be avoided either in the bulk of the material (i.e., hole-trapping impurity levels <<0.1% are typically required) or at interfaces. Another function of the HTL is that it should act as an electron-blocking layer to prevent the flow of electrons from the EML and ultimately to the anode. For this purpose, a very shallow LUMO level is desirable.

Materials having low ionization potential together with low electron affinities and high hole mobility usually function as hole-transporting materials by accepting and transporting hole carriers. The most common hole-transport materials are N,N'-diphenyl-N,N'-bis(3-methylphenyl)1,10-biphenyl-4,40-diamine (TPD), N,N'-diphenyl-N,N'-bis(1-naphthylphenyl)-1,10-biphenyl-4,40-diamine (NPB), and 1,10-bis(di-4-tolylaminophenyl) cyclohexane (TAPC). Ongoing efforts on the development of HTLs include the improvement of thermal and electrochemical stability, mobility, glass transition temperature, and reduction in the energy barrier interface between the anode and HTL and the crystallization behavior.

The ETL functions as a conducting material to help transport electrons from the cathode and into the organic layers of the device—ideally, transporting the electrons via a hopping mechanism involves transitory production of anion radicals (negative polarons) in the molecules involved. As such, the material needs to have a LUMO level close in energy to the work function of the cathode material used so as to aid charge injection. It also needs to be comprised of a material that is relatively stable in its one-electron reduced form. As with all organic layers, it should form good amorphous films and have a high transition temperature to favor stable operation over extended periods.

Since most of the high-efficiency organic emitters have p-type character and mainly hole-transporting behavior, to achieve high efficiency device performance an electron-transport layer is necessary to balance the charge injection and transport. In fact, it is documented that introducing an ETL into OLEDs results in orders of magnitude improvement in the device performance. The functions of the ETLs are to reduce the energy barrier between the cathode and the emitter and to help the electrons to be easily transported to the emitter. The introduction of an ETL lowers the energy barrier between the LUMO level of the EML and the work function of the cathode for electron injection. Meanwhile, most ETLs also serve as hole-blocking layers to efficiently confine the exciton formation in the EML and thus balance charge injection. It also prevents the charge leakage and the accumulation of charges at the cathode and ETL interface.

Materials having good electron-transporting and hole-blocking properties (i.e., electron mobility higher than $10^{-5}\,cm^2/Vs$) and high electron affinities together with high ionization potentials are the favorite materials for accepting negative charges and allowing them to move through the molecules. The most common ETL materials are aluminum-tris-8-hydroxyquinoline (Alq_3) and 9,10-di(2-napthyl)anthracene (ADN).

Finally, let us consider in more detail the EML, which is considered the distinctive layer in an OLED device. Indeed, the major part of the molecular design and engineering of materials comprising OLEDs is devoted to the emissive materials. In many cases, however, the EML is actually a mixture of two or more materials wherein there is at least one electroluminescent emissive material coupled with a charge-transporting host material. Such guest–host systems are extremely common in OLEDs based on small molecules (SMOLEDs, small-molecule organic light-emitting diodes), whereas in polymeric OLED devices, the emitter layer is usually composed of a single polymer (PLEDs, polymer light-emitting diodes), which combines both the light formation and charge-transport functionalities into a single-phase material. Clearly, this is a broad generalization given that it is possible to use a single material for the emitter layer in SMOLEDs and multiple-phase layers (e.g., polymer blends or doped polymers) in polymeric OLEDs.

In general, SMOLEDs contain small-molecule emissive materials that can be processed by either vacuum deposition techniques or solution coating. The emissive small molecule may be a fluorescent (singlet excited state) or a phosphorescent (triplet excited state) emitter.

PLEDs contain polymeric emissive materials that are almost exclusively processed by solution coating. In general, fluorescent emission is observed in PLEDs, but there are only few examples of phosphorescent materials being incorporated into a polymer chain and used as phosphorescent emitters.

In spite of technological issues of efficiency and stability of PLEDs as compared to SMOLEDs, the former promises to revolutionize the display-manufacturing technology as it provides the possibility of inexpensive solution fabrication.

Indeed, ambient temperature and pressure fabrication conditions (spin coating, bar coating, inkjet printing, etc.) of PLED-based large-area screens, enabled by good film-forming properties of polymers, are particularly attractive for the industrial application. However, purely polymer-based LEDs present external quantum efficiency (EQE) of less than 10%, which limits the achieved power efficiency below ~20 lm/W. Besides the energy consumption issue, low efficiency also poses a problem of heat dissipation, which affects the device stability.

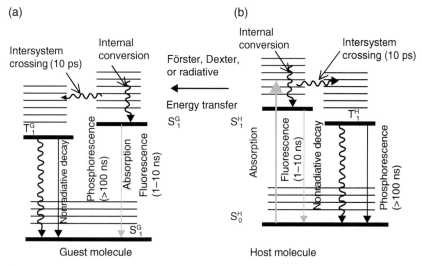

FIGURE 2.2 (a, b) Electronic processes of host–guest molecules, in which the guest molecules can emit light through both singlet and triplet states. S_0^H, S_0^G: a singlet ground state of the host and guest molecules, respectively. S_1^H, S_1^G: a first excited singlet state of the host and guest molecules, respectively. T_1^H, T_1^G: a first excited triplet state of the host and guest molecules, respectively.

In such highly complex, multilayer, and multicomponent OLED devices, a successful strategy to effectively master the distribution of the excitation in the desired emitting molecules is to manage the dynamics of the various energy-transfer mechanisms taking place in the luminescent active layers. When a host molecule in the typical host–guest EML is excited from the ground state by either absorbing light energy or being driven by electric energy to a higher vibrational energy level, it is subjected to collisions with the surrounding molecules. It can directly release its energy through radiative decay or nonradiative decay processes to the ground state, or in the presence of a suitable guest molecule, energy-transfer processes may occur. The latter event, depicted in (a) of the diagram reported in Figure 2.2 as an energy-transfer transition from the host molecule to the guest molecule, occurs through Förster, Dexter, or radiative energy-transfer processes. At this point, the radiative decay processes will occur from the luminescent guest molecules. It may be noted that the emission spectrum observed is sometimes the emission from the guest molecules only due to complete energy-transfer processes, but sometimes it combines the guest and host molecule emission due to incomplete energy transfer (Figure 2.2).

Because molecular excited states may also transfer from molecule to molecule while conserving their spin and energy, one can treat them as quasiparticles named excitons. The highly localized excited states are known as Frenkel excitons, having radii of a few angstroms.

One can treat the Frenkel exciton formation as due to the hop of charge carriers (electron, hole) to a neighboring molecule. Due to the fact that the rate of exciton hopping is given by the multiplication between the rate of electron transfer and the rate of hole transfer, the theory of electron transfer can shed light on the understanding of exciton hopping.

During OLED operation, singlet and, in some cases, triplet excitations may first be created in the host material. Then, through charge or energy transfer from the host to the guest, singlet or triplet excited states are formed in the guest. For an effective host-guest system, several factors have to be considered, such as the phase compatibility of the host and guest, the aggregation of the molecules, and the host–guest energy level alignment.

In this charge/energy-transfer process, the band gap of the guest should fall within the band gap of the host to favor transport of electrons and holes from the host to the guest, where they should then recombine (Figure 2.2). In order to dominate efficient energy-transfer process requires that the energy of the excited state of the host should be higher than that of the emissive excited state of the guest. This applies to both singlet excited states and triplet excited states of the host and the guest as shown in Figure 2.2.

The efficiency of energy transfer for the singlet excited (fluorescent) state is easy to verify if there is an overlap between the emission spectrum of the host and the absorption spectrum of the guest. Beyond this requirement, for an efficient energy transfer from the host to the guest of the triplet state (phosphorescent), the excited triplet state of the host should be higher than that of the guest.

Finally, we have to mention the other photophysical mechanism, which is typically used for localizing the excitons on the guest molecules. In the *charge-trapping* process, a hole (electron) generated in the host during device operation is directly localized on the guest molecule if the HOMO (LUMO) of the guest lies above (below) that of the host material. Then, the counterparticle is trapped on the guest energy well, thus forming an exciton. The higher the difference between the HOMO (LUMO) levels of the host and the guest, the higher is the efficiency of hole (electron) trapping, and sometimes, direct charge trapping could be the prime mechanism of the exciton generation on the guest molecules. However, the charge trapping also creates a barrier for charge transport across the device, resulting in significant increase of the operating voltage.

In general, the approach of engineering the active stack in the EML is fundamental in improving the internal efficiency of OLEDs. Indeed, the majority of organic semiconductors form amorphous, disordered films [10]. As a consequence, charges are injected statistically with respect to their electron spin, finally determining the formation of singlet and triplet excited states. Because the triplet state has a multiplicity of 3 [10], on average 75% of the excitons formed are triplet states, with the remaining 25% being singlets. Segal *et al.* [11] observed slightly smaller values for the singlet fraction in both small-molecule and polymeric systems [(20% ± 1%) and (20% ± 4%), respectively], which are in rather good agreement with this simple statistical picture.

The low singlet fraction causes OLEDs based on fluorescent emitter molecules to be rather inefficient with an upper limit of the internal quantum efficiency $\eta_{int} = 25\%$, because emission solely occurs in its singlet manifold.

Several routes have been proposed to obtain a higher η_{int} through the efficient harvesting of excitons in OLEDs, in particular triplet excitons. Historically, the very first experiments with ketone derivatives which showed intense phosphorescence at low temperature opened a new method for triplet harvesting even though the EQE (η_{EQE}) was limited to a low value [12]. Rare metal complexes containing Eu and Tb established intramolecular cascade energy transfer as another route to harvest triplet excitons but did not show promising η_{EQE} [13–15]. Later, a successful strategy was realized using room-temperature phosphorescent emitters such as platinum and iridium complexes. In this case, following the mixing of the spin orbitals of S_1 and T_1 states due to the presence of a heavy atom, the radiative decay rate from the T_1 state to the ground state is significantly improved.

Along with realizing a highly efficient emissive triplet state in a molecule, the heavy-metal effect strongly enhances the intersystem crossing rates between the singlet and triplet manifolds [16]. Thus, the fractions of singlet excitons that are created under electrical excitation are efficiently converted into triplet states before they can recombine radiatively. The intersystem crossing rate is close to unity in various phosphorescent systems [17]. Therefore, phosphorescent materials in OLEDs can lead to internal EL efficiencies of $\eta_{int} = 100\%$.

Furthermore, state-of-the-art emitters are especially optimized for having short excited-state lifetimes in order to reduce bimolecular quenching processes, which limit the photoluminescence quantum yield at high excitation levels [18,19].

In addition, the utilization of phosphorescence emitters as a triplet sensitizer has been proposed [11,12]. Using this process, triplet harvesting realized by energy transfer from the T_1 state of a phosphorescent emitter such as an iridium 2-phenylpyridine complex to the S_1 state of a fluorescent emitter via dipole–dipole coupling (i.e., Förster energy transfer) resulted in an

$\eta_{int} = 45\%$ [20]. However, the rather limited η_{int} is due to the presence of the competitive deactivation process of triplet–triplet energy transfer.

Although OLEDs based on fluorescent molecules, which are composed of simple aromatic compounds, have continued to attract interest because of their longer operational lifetimes in the blue-emitting range, higher color purity (narrow spectral width) EL, and broader freedom of molecular design compared with phosphorescence-based OLEDs [21–23], the η_{int} of traditional fluorescence-based OLEDs is limited to less than 25% even in the ideal case. Therefore, the enhancement of η_{int} in OLEDs using conventional fluorescence-based emitters is still obviously a major concern for the development of future OLEDs.

A concept to improve the internal quantum efficiency of fluorescent EL makes use of the high triplet density via delayed fluorescence [10]. Here, the interaction of two triplet states (called triplet–triplet annihilation) will create delayed singlet excitons: $T_1 + T_1 \rightarrow S_0 + S_n$ [24]. Based on this nonlinear process, an internal electron–photon conversion efficiency much higher than expected is reached. The device data of Okumoto et al. [25] showing a twofold improvement to the $\eta_{int} = 25\%$ limit (nearly 10% EQE) suggest that this process takes place.

Kondakov [26] gave experimental evidence that delayed fluorescence substantially contributes to the internal efficiency of fluorescent OLEDs. Endo et al. [27] suggested an alternative concept [thermally activated delayed fluorescence (TADF)] to feed the singlet state of a molecule with its triplet excitons.

The typical energy diagram of a conventional organic molecule, depicting singlet (S_1) and triplet (T_1) excited states and a ground state (S_0) is reported in Figure 2.3 [28]. In standard state-of-the-art phosphorescent systems, the S_1 level is considerably higher in energy than the T_1 level, by 0.5–1.0 eV, because of the electron exchange energy between these levels (Figure 2.3). However, a careful design of organic molecules can lead to a small energy gap (ΔE_{ST}) between S_1 and T_1 levels. Correspondingly, a molecule with efficient TADF requires a very small ΔE_{ST} between its S_1 and T_1 excited states, which enhances $T_1 \rightarrow S_1$ reverse intersystem crossing (RISC). Such excited states are attainable by intramolecular charge transfer within systems containing spatially separated donor and acceptor moieties [28]. The critical point of this molecular design is the combination of a small ΔE_{ST} of ~100 meV with a reasonable radiative decay rate, to overcome competitive nonradiative decay pathways, leading to highly luminescent TADF materials. Because these two properties conflict with each other, the overlap of the HOMO and the LUMO needs to be carefully balanced. Furthermore, to enhance the photoluminescence efficiency of a TADF material, the geometrical change in molecular

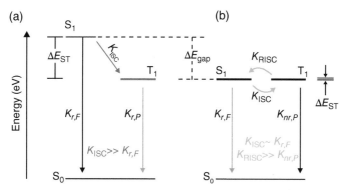

FIGURE 2.3 Comparison between the (simplified) energy level diagram for standard phosphorescent (a) and TADF-based (b) emitters. Important rates k_i are indicated. r, radiative; nr, nonradiative; F, fluorescence; P, phosphorescence; ISC, intersystem crossing; RISC, reverse ISC; ΔE_{ST}, singlet–triplet splitting. [*Source*: Reineke [28], Figure 1, p. 269. Adapted with permission from the Nature Publishing Group.]

conformation between its S_0 and S_1 states should be restrained to suppress nonradiative decay. The RISC rate (k_{RISC}) (Figure 2.3), which is the rate-limiting step in TADF emitters, has been demonstrated to be as high as $10^6\,s^{-1}$. As shown in Figure 2.3, state-of-the-art phosphorescent molecules possess radiative rate constants of the same order of magnitude.

Devices based on this concept showed a very high RISC efficiency of 86.5% and an EQE beyond the fluorescence limit of 5%. Recently, Uoyama *et al.* [29] reported promising OLED performance data based on this TADF concept. With a specially designed novel class of organic materials, the exchange splitting could be reduced to approximately 80 meV, giving rise to an effective RISC. These materials possess a very high rate of delayed fluorescence, which is comparable to the radiative rates of phosphorescent emitters [30]. In their report, OLEDs are discussed reaching 19% η_{EQE}, which is in line with the currently used phosphorescent emitter technology. It is expected that TADF will potentially allow internal quantum efficiencies of light emission of 100%, similarly to phosphorescence allowing for the development of truly spin-indifferent organic LEDs.

Although the scientific community is actively pursuing research into this typology of materials, the industry may be reluctant to embrace TADF-based OLED concept immediately. At the production level, existing OLED products have resulted from years of development and radical new designs introduce risk, substantial investment, and facility downtime. The lower cost of TADF emitter synthesis relative to the high cost of transition metals such as iridium may alone be insufficient to persuade the industry to make a technology switch. However, if TADF does lead to blue OLEDs with considerably superior stability compared to phosphorescence, the industry is likely to consider adopting it.

We have also to mention that Segal *et al.* [31] introduced the mixing of charge-transfer (CT) states in small-molecule host–guest system for enhancing the emission from fluorescent material (extrafluorescence). By assuming that exciton formation follows the relaxation of a CT state consisting of an electron and a hole on neighboring molecules, the anomalous higher lying triplet CT state can be exploited for increasing the rates of singlet exciton formation, leading to a singlet fraction as high as $\chi_S = 0.84 \pm 0.03$ [31].

Aside from the optimization of the device multistack architecture and the photophysical processes ruling the light formation, a major limiting factor in terms of efficiency is the outcoupling of light [32]. Since a standard OLED consists of a highly reflective cathode on one end and a semitransparent anode on the opposite end, it essentially forms a microcavity where certain modes (or wavelengths) of light are enhanced whereas other modes are trapped inside, depending on the total device layer thickness [33]. This is related to the fact that standard organic materials as well as the indium tin oxide (ITO) in the anode have higher refractive index values (~1.7–1.9) than those of the glass substrate (~1.51) and air. This results in even more modes of light being trapped in the high refractive index layers in the device due to internal total reflection and consequentially leads to a directional (angular-dependent) emission profile [34].

We consider in more detail the various light propagation modes in a conventional bottom-emission OLED (see the device cross section reported in Figure 2.4). They are mainly determined by the thin-film structure of the device and the respective optical properties (i.e., refractive indices and absorption coefficients) [38].

In general, the generated photons can outcouple to an external light mode (the so-called far-field, air, or outcoupled modes) and leave the device through the transparent anode (ITO) and substrate (glass). This, however, is only accomplished by 20–30% of the photons generated within an emission angle cone of around 40°. The majority of the photons outcouple to either substrate modes in the glass or waveguide modes in the ITO and organic layers.

Coupling to organic or waveguide modes [35] (Figure 2.4) occurs when the photon path exceeds the critical angle of total internal reflection due to the large refractive-index mismatch between the organic layers, substrate, and air. In the first approximation, two optical interfaces, that is, the organic/substrate and the substrate/air interfaces, are formed, where total internal reflection may occur. Coupling reflects the photons back through the substrate

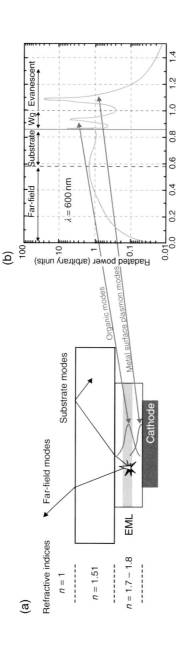

FIGURE 2.4 (a) Cross section of an OLED with indication of different light modes. (b) Typical power spectrum of the internally generated light shown as a function of the in-plane wavevector. Vertical lines separate the various possible light modes. Neither waveguide (wg) nor evanescent modes (thick line) can be accessed with external-light outcoupling techniques; thus, they dissipate within the layer structure. Model calculation [36] for a bottom-emitting device similar to devices discussed in Reference [37]. [*Source:* Reineke *et al.* [35], p. 1252. Adapted with permission from American Physical Society.]

and organic layers to the reflective cathode. The segmented path taken by the photons is well described by geometrical ray optics and results in traveling distances in the centimeter range.

The waveguide modes trapped in the organic-transparent electrode layer system can be one of two types: either a zigzag mode that has a maximum in the organic layer or a plasmonic mode that propagates mainly along the highly reflective metal cathode and is therefore quickly absorbed [39–41]. Unfortunately, most of the photons that enter into a waveguide mode end up in the plasmonic mode given that this process is very efficient for short distances between the EML and the cathode.

Moreover, Figure 2.4 additionally shows a power spectrum, obtained from model calculations [36] of a conventional monochrome bottom-emitting OLED [32,42], plotted as a function of the in-plane wavevector [32,36,42]. In such a power spectrum, the modes discussed can easily be attributed to different ranges of the in-plane wavevector, indicated by the vertical lines in Figure 2.4. Here, the fraction of photons that directly leaves the device (far field) typically is in the range of only 20% [36,37,39,43,44]. More light can be extracted to the far field by applying modifications of the substrate/air interface by converting substrate into air modes [38,45–47,108]. On the contrary, as indicated by the thick solid line, modes with larger in-plane wavevector, that is, waveguide and evanescent surface plasmon modes, cannot be outcoupled by external techniques.

Different techniques in internal or external device modification are usually implemented in order to enhance light outcoupling efficiency in OLEDs. One common approach is to adopt thin-film outcoupling techniques by making use of refractive index modulation layers [48,49]. However, these techniques are typically effective for only a narrow range of emission wavelengths, sufficient for a single color OLED, but not for a white OLED with a broad-band emission spectrum.

Another popular approach is to use photonic periodic structures such as Bragg gratings [50,51] and low-index grids [46] in the device, which again only allows light management for certain modes of light. Moreover, these methods could not effectively address the issue of angular dependence in the emission profile. A more feasible outcoupling approach for OLED is to employ a buckling structure where a thick, corrugated architecture with a broad periodicity distribution persists throughout the entire OLED, such that a wide range of modes are enhanced and minimal angular dependence of the emission profile is present [52]. This technique can be readily combined with a number of cost-effective substrate surface texturing techniques such as the incorporation of an array of microlens [53] or micropyramids [47] to extract even more modes of light out of the device. Although this technique does

(a)

Plastic
Ta₂O₅
Au
Organic
Al

(b)

FIGURE 2.5 (a) Schematic of the OLED device structure on low-cost flexible plastic with metal electrodes. (b) Photograph of a large-area flexible OLED ($50 \times 50\,mm^2$) working at high luminance ($>5000\,cd/m^2$). [*Source*: Wang *et al.* [48], Figure 1, p. 754. Adapted with permission from Nature Publishing Group.]

improve outcoupling performance [35], it provides only a small enhancement to a mediocre OLED and is therefore incapable of reaching the efficiency levels attained by the best OLEDs.

The use of high-refractive-index substrates [47] that are better matched to that of ITO is also effective but less practical since high-refractive-index substrates are much more expensive than most commercial glass and flexible plastic substrates. Even though the plasmonic modes still exist but can be well suppressed, as recently shown for high-efficiency white OLEDs [54].

In particular, enhancements in optical outcoupling that rely on the use of high-refractive index layer is not recommended for low-cost mass production of OLEDs, which implement lightweight, flexible plastic substrates, because most plastics have a low refractive index ($n \leq 1.6$) that is comparable to standard glass. Thus, new optical outcoupling strategies are required to realize high-performance OLEDs on flexible plastic for the next generation of mass-produced flexible displays and solid-state lighting.

Wang et al. [48] presented a new paradigm for the optical outcoupling enhancement of OLEDs that is fully compatible with flexible plastic substrates with low refractive index. The key step of their technique is to replace the ITO transparent electrode with an oxide–metal–oxide electrode stack. They employed a multifunctional anode stack consisting of a thin semitransparent gold layer, which serves as a conductive electrode, sandwiched between a thin-film high-refractive-index layer made of tantalum oxide (Ta_2O_5) on a flexible plastic substrate—the optical coupling layer— and a hole-injection molybdenum trioxide organic layer. The gold layer forms a weak optical microcavity with the aluminum cathode. Because the design exploits a plastic substrate with a relatively low refractive index (<1.6), a high-refractive-index glass substrate is not required (Figure 2.5).

A record high EQE of ~40% at a very high brightness of 10,000 cd/m² was achieved using this new electrode design for a green OLED fabricated on flexible plastic (Figure 2.5). Additionally, after further reduction of the amount of light trapped in the plastic substrate using a lens-based structure on top of the device, the EQE and power efficiency at 10,000 cd/m² are increased to 60% and 126 lm/W, respectively.

2.2 APPLICATIONS OF OLED TECHNOLOGY

Tremendous progress has been made on OLEDs in the past two decades [47,48,54–58]. Since Kodak developed the first low-voltage OLED using a simple bilayer structure [59], Kido group [54] has demonstrated the first white OLED by mixing different colored emitters. Forrest's group [55] later introduced phosphorescent emitters to quadruple the device efficiency by harvesting both singlet and triplet excitons through efficient intersystem crossing activated by the presence of the heavy metal in the emitter. Shortly after, Thompson's group [60] developed a platform for the synthesis of currently predominant Ir-based phosphors.

Recently, Leo group [47] reported an improved OLED structure, which reaches white fluorescent tube efficiency by combining a carefully engineered emitter layer with high-refractive-index substrates and using a periodic out-coupling structure. Particularly, it achieved a device power efficiency of 90 lm/W at 1000 Cd/m^2.

Indeed, OLEDs have high color quality and, as they can be made into large sheets, they can replace fluorescent lights that are currently used in houses and buildings with potentially reduced energy costs for lighting. Full-color capability is an essential feature in organic electroluminescent devices for flat-panel display. Moreover, OLED technology looks very promising for revolutionizing eco-sustainability in solid-state lighting in the next decade. This energy-efficient lighting technology may play an important role in reducing global consumption of electricity by almost 50%.

It is worth mentioning the synergetic strategy implemented by Leo group in order to engineer OLED devices capable of outperforming fluorescence tubes. The key feature of the white OLED layer structure is the positioning of the blue phosphor within the EML and its combination with a carefully chosen host material. Indeed, it is well known that for power-efficient white OLEDs, the high-energy blue phosphors demand host materials with even higher triplet energies to confine the excitation to the emitter [60]. Taking exciton binding energy and singlet–triplet splitting into account, the use of such host materials considerably increases the transport energy gap and therefore the operating voltage. For these reasons, blue fluorescent emitters are widely used to complete the residual phosphor-based emission spectrum [61,62]; this, as we have already said, either reduces the internal quantum efficiency or requires blue emitters with special properties.

A novel concept for achieving energy-efficient photon generation consists in locating the triplet energy level of the blue emitter material in resonance with its host so that the blue phosphorescence is not accompanied by internal triplet energy relaxation before emission. The exciton formation region is at the interface of a double-emission-layer structure [63] as reported in the energy level diagram of Figure 2.6. The blue host–guest system is surrounded by red and green sublayers of the EML to harvest unused excitons. Nonetheless, the different sublayers are separated by thin intrinsic interlayers of the corresponding host material to decouple the sublayers from unwanted energy transfer. In particular, the introduction of high-triplet-energy interlayer

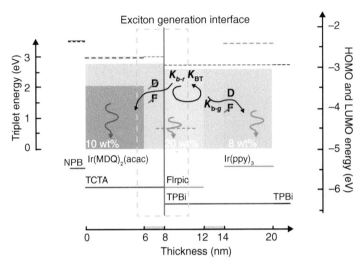

FIGURE 2.6 Energy level diagram in an OLED reported in Reference 47. Lines correspond to HOMO (solid) and LUMO (dashed) energies; filled boxes refer to the triplet energies. The orange color marks intrinsic regions of the emission layer. F and D represent Förster- and Dexter-type energy exchange channels, respectively. The orange dashed box depicts the main region of exciton generation. [*Source*: Reineke *et al.* [47], Figure 1, p. 235. Adapted with permission from Nature Publishing Group.]

tris(4-carbazoyl-9-yl-phenyl)amine (TCTA) strongly inhibited the exciton migration (by either Förster or Dexter diffusion processes) from the blue to the red emitters. Furthermore, the location of the triplet energy levels in resonance between the blue emitter and its host (TBi:FIrpic) allows for back energy-transfer process from the guest sites to the host with consequent delayed component in the decay of the emitting species. Finally, efficient green emission is achieved by diffusion process to green emitters, that is, Ir(ppy)$_3$, by exciton harvesting due to energy-exchange-driven diffusion.

Finally, the modification of the substrate is implemented in order to achieve efficiencies as high as 90 lm/W. Particularly, glass modes are converted into far-field modes by using patterned high-refractive-index glass for vanishing the index mismatch between organic materials and substrate.

However, the lifetime issue of the blue emitters has to be solved and the fabrication cost has to be significantly reduced for a broad application in general lighting.

During the past decade, traditional LED technology has also made incredible progress in improving the energy conversion efficiency, in

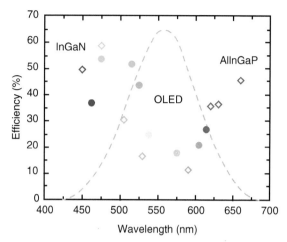

FIGURE 2.7 Current status of energy conversion efficiency of OLEDs (in solid circles) and LEDs (in open rhombuses) in the visible spectrum. The LED data were taken from [64]. The OLED efficiencies are power efficiencies of the device at $1000\,cd/m^2$ after applying the outcoupling enhancement technique listed in Reference 48 and normalized to the theoretical limit for the corresponding wavelength. The OLED data were taken from References [37,48,57,65–70]. The dashed gray curve represents the photopic sensitivity response curve of the human eyes. [*Source*: Chang and Lu [71], Figure 2, p. 460. Adapted with permission from IEEE.]

particular using GaN-based blue LEDs. Indeed, traditional LED technology has been problematic in producing LEDs in the green-yellow color range, the so-called "Green-Yellow" gap issue. This is unfortunate as the human eyes have the highest sensitivity to these wavelengths. OLEDs, however, face no such constraints. Figure 2.7 shows the efficiency of the state-of-the-art OLEDs [37,48,57,65–70] compared to state-of-the-art inorganic LEDs [64]. Even though GaN-based LEDs are best performing in terms of conversion efficiency in the blue emission, OLEDs are superior in the range of the visible spectrum that is most sensitive to the human eye, particularly in the green portion. However, there is still much room for improvement to be made on yellow, orange, and red OLEDs. In fact, from the energy gap law [69], it is anticipated that the efficiency of yellow and orange OLEDs can surpass that of the red OLEDs. Currently, the blue OLEDs are closely catching up in terms of efficiency compared to inorganic LEDs; however, they are well known to have rather short device lifetime due to inherent instability of the blue phosphors, leading to a rapid efficiency drop at higher luminance levels needed for lighting [47].

Indeed, real lighting applications specifically require device brightness level around 1000 cd/m^2 (up to 5000) while preserving high internal quantum efficiency. Initially, 1000 cd/m^2 value was established in the literature as a level to ensure best device comparability. Nonetheless, in the last few years, 3000 cd/m^2 was increasingly quoted as the standard level for OLED lighting applications.

As we have already stated, the quantum efficiency in phosphorescent systems is subjected to noticeably drop at high brightness (efficiency roll-off) due to intrinsic long electronic lifetime of the radiative emissive species.

Even though state-of-the-art phosphors have excited-state lifetimes down to 1 μs, the lifetime is still about orders of magnitude longer than their fluorescent counterparts, which is the main reason for different electroluminescent properties of fluorescent and phosphorescent emitters [10]. The brightness L of an OLED is proportional to the excited-state exciton density n and inversely proportional to the excited-state lifetime τ [19]: $L \sim n/\tau$. Considering the spin statistics, one derives for a fluorescent (fl) and a phosphorescent (ph) system, respectively:

$$L \sim \frac{n_{fl}}{4\tau_{fl}} \qquad L \sim \frac{n_{ph}}{\tau_{ph}}$$

This leads to an expression for the ratio between the excited-state densities n_{ph}/n_{fl}:

$$\frac{n_{ph}}{n_{fl}} \sim \frac{\tau_{ph}}{4\tau_{fl}} = \left[\frac{1\mu s}{4 \times 10\,ns} = \frac{25}{1} \right]$$

when considering representative fluorescent and phosphorescent excited-state lifetimes set to 10 ns and 1 μs, respectively [19,72].

The direct consequence is illustrated in Figure 2.8. To reach the same luminance level in fluorescent and phosphorescent systems, the exciton density is typically about 25-fold higher in the case of phosphorescence, which increases the probability of excited-state annihilation processes, such as triplet–triplet annihilation, triplet–polaron quenching [18,19], and, in some cases, field-induced exciton dissociation [73].

Despite the higher luminous efficiency at higher brightness, it is unlikely that fluorescence-based OLEDs will be able to compete with phosphorescence-based OLEDs, given the approximately 75% recombination losses [11] within the device.

In general, achieving the highest possible device efficiencies needs the sophisticated, sometimes highly complex, device layouts that we have briefly

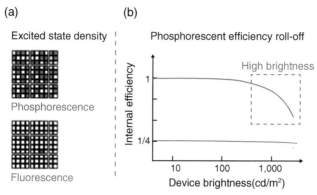

FIGURE 2.8 (a) The excited-state density in the cases of phosphorescence and fluorescence, respectively. (b) Typical efficiency versus brightness characteristics. In the case of phosphorescence, the efficiency drastically decreases at high brightness as a consequence of quenching processes. [*Source*: Reineke *et al.* [35], p. 1252. Adapted with permission from American Physical Society.]

introduced in the previous paragraph. These designs are not desirable for upscaling the device production to reasonable OLED panel sizes. Thus, devices offering high-efficiency and large-area controllable device architectures are the concepts of choice.

Even though the first reports on triplet harvesting employed rather complex EML designs [63], the layer complexity has recently been greatly reduced. Rosenow *et al.* [74] introduced a triplet-harvesting EML consisting of two simple 5-nm-thick sublayers, offering great reproducibility. Another attractive concept promising high efficiency is the combined monomer–excimer phosphorescence [75]. Especially, the possibility to design a white OLED based on a single emitter that is dispersed into a matrix material at a certain concentration offers unmet simplicity.

This approach can reach similar high EQE values than conventional phosphorescent devices. To date, their corresponding luminous efficacies are smaller than the corresponding values of other concepts. This is mainly due to the superior electrical performance of the respective devices [47,65,70].

Moreover, also the balance between the light emission and brightness has to be suitably optimized in order to achieve a luminous flux in OLEDs that matches the output of an incandescent bulb. Interestingly, even large OLED areas of $50 \times 50\,cm^2$ need a luminance of $680\,cd/m^2$ (about a factor of 2–3 brighter than a typical computer display) to reach the flux of the light bulb. As we have already observed, the high brightness operation of OLEDs is accompanied with a decrease in device efficiency as a consequence of excited-state annihilation processes [8,19,73]. Furthermore, the device long-term stability is inversely proportional to its operating brightness [76–79]. On the other hand, the production costs of an OLED panel, and therefore the costs per lumen, increase roughly linearly with the panel area.

However, apart from the many hurdles to overcome in order to improve device optoelectronic performances, OLED technology evidently offers several advantages over conventional lighting technologies. OLEDs can provide revolutionary lamp properties, which include their excellent transparency, color tunability, and flexible lighting sources. Particularly, the thickness of the OLED panel for lighting applications reaches just 1 mm. Such a characteristic may allow OLED lighting to be placed directly on ceilings rather than using fixtures that are suspended from them. In addition, OLED's flexibility may allow lighting fixtures to operate effectively when designed for spaces with severe constrains.

Moreover, the different manufacturing processes of OLEDs have the potential to offer, in principle, many advantages over traditional flat-panel displays. OLEDs can be printed onto a substrate using traditional inkjet technology, which can be cost competitive with respect to liquid crystals for large-area displays. OLED screens provide clear and distinct image, even in bright light, and they can be viewed from almost any angle up to 160° [80]. OLED displays produce good brightness and clarity and consistent image quality with good contrast and high luminescence efficiency. Unlike liquid crystal displays (LCDs), they have neither backlights nor chemical shutters that must open or close. OLED pixels turn on and off as fast as any light bulb and hence have fast response time. Thin OLED screens are free from the added bulk and weight of backlighting and are tough enough to use in portable devices such as cell phones, digital video cameras, and DVD players. Indeed, the inherent structural and functional properties of organic light-emitting devices based on small-molecules or polymers OLEDs are virtually ideal for flat-panel display (FPD) applications. Particularly, the characteristics of OLEDs that allow for the implementation of an organic FPD technology are the following: Lambertian self-emission [81], which produces a wide viewing angle; fast response time (below microseconds), which is suitable for moving images; high luminous efficiency and low operation voltage, which guarantee low power consumption; lightweight, very thin structure, and robustness to the external impact as well as good daylight visibility through high brightness and contrast, which are desirable for portable displays; broad color gamut, which is suitable for full-color displays; thin film conformability on plastic substrates.

However, the practical implementation of OLEDs (or organic thin films in general) in modern electronic circuits will ultimately be decided by the ability to produce devices and circuits at a cost that is significantly below that

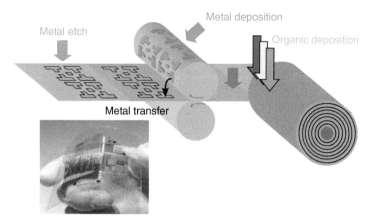

FIGURE 2.9 Conceptual diagram of continuous and very low cost manufacture of organic electronic devices. [*Source*: Forrest [5], Figure 7, p. 916. Adapted with permission from Nature Publishing Group.]

needed to manufacture conventional electronic circuits based on silicon. If successful, the low-cost fabrication processes will ultimately result in the "printing" of large-area organic electronic circuits using roll-to-roll methods, where low-temperature deposition of the organics is followed by metal deposition and patterning in a continuous, high-speed process. In Figure 2.9, a conceptual diagram of continuous and very low cost manufacture of organic electronic devices is reported. Starting with a roll of plastic substrate, the organic materials forming the device active regions are first deposited from either the liquid or vapor phase, and then, an unpatterned metal layer that works as "strike layer" is deposited. The film is then passed between rollers with an embossed pattern representing the ultimate electrode scheme required on their surfaces. These rollers then directly pattern the electrodes. In the final step, the strike layer metal is removed by dry etching, completing the circuit. The sheets are eventually coated to eliminate degradation from exposure to the environment and then packaged to suit the desired application.

Even though very few of these technologies have left the laboratory and found their way into a manufacturing environment for fabricating high-performing and long-lifetime thin-film organic devices, it is evident that the inherent structural characteristics of organics opened up the door to a suite of innovative fabrication methods that are simpler to implement on a large scale than what was ever thought possible in the world of inorganic semiconductors [71].

Once the inexpensive and reliable print-based manufacturing chain is developed for the production of light-emitting displays based on organic devices, numerous new market opportunities will be enabled, given the OLEDs distinguishing features such as low operating voltage requirements (2–10 V), inexpensive DC driver, electronics, moderate lifetimes (>10,000 h), full red, green, and blue (RGB) color range, and potential for very high brightness (>2000 cd/m^2) that enables passive addressing and viewing under daylight conditions.

The market opportunities include electronic displays used in high-resolution displays, such as those found in laptops and personal digital assistants (PDAs), as well as displays found in the low-resolution display market (defined as displays having less than 100,000 pixels), such as alphanumeric displays. Being able to operate in both the printing and electronic display industries, significant untapped target markets exist for emerging organic display technology. For example, most high-resolution display manufacturers are unable to meet the needs of the in-store media market because the market requires low-cost electronic displays with peel-and-stick qualities, as it is in the case of practically disposable alphanumeric displays that could be integrated into temporary and semipermanent signage. Additionally, inexpensive display technology could quickly impact the industrial and electronic market with a myriad of printed alphanumeric, segmented, and pixilated displays. In time, fully printed organic light-emitting displays may penetrate the billion-dollar market for large format LED displays, as well as the multibillion dollar market for print-on-vinyl displays used throughout the outdoor advertising world, by delivering dynamic content to public environments [81].

However, many obstacles must be overcome before the potential of OLED technology can be realized. These include operating and storage lifetime, environment sensitivity to water and oxygen, drive schemes (passive- vs active-matrix approach), power efficiency and consumption; integration with the drive electronics, fine patterns control with vivid colors, high light output extraction, and the full cost analysis of manufacturing (see Chapter 7 for an extended discussion on organic electroluminescence display technology).

Focusing on developing a fabrication protocol for producing large-area flexible liquid-based (or hybrid-based), organic light-emitting displays, three different technical issues are to be resolved. Firstly, an organic electro-luminescent (EL) ink ought to be developed, which is suitable for patterning in a uniform and repetitive manner, by using a printing process that is scalable to high-volume manufacturing. Secondly, develop a method for patterning every layer in the device using a print-based process ought to be engineered. Thirdly, an encapsulation process has to be optimized capable of preserving the mechanical flexibility of the substrate and guaranteeing a sufficient cost-effective device lifetime. Moreover, in active matrix OLEDs where each pixel is powered by a transistor, the light generated can escape either between the transistors (anode side or "bottom emission") or, by making the cathode transparent, through the cathode ("top emission"). The advantage of the bottom-emitting device is that it is a simpler structure. The disadvantage is that the transistor blocks a significant proportion of the light and the pixels themselves have to be brighter to compensate the light loss with consequent increase in aging and shortening of lifetime of the device. In the top-emitting structure, the proportion of generated light that reaches the viewing eye is significantly greater, enhancing efficiency and, because the pixel can be run less bright, extending lifetime. However, the device requires a cathode of high transmissivity and a reflecting anode (exactly the reverse with respect to bottom-emitting device with a reflecting cathode and a transparent anode).

To solve some of these important issues, tremendous research efforts from academia and industry have been underway since the first appearance in 1997 of the OLED-based monochromatic car stereo displays in the market [82]. AM-OLEDs based on thin-film transistor (TFT) technology, in particular, have attracted considerable attentions for high-resolution, large-size FPD applications, such as laptop and TV screens. Several FPD companies (such as Toshiba and Matsushita, Kodak and Sanyo, Sony, Samsung SDI, Chi Mei Optoelectronics, and IBM Japan) [83,84] have reported 15- to 24-in. TFT AM-OLED prototypes with wide extended graphics array WXGA (1200×768) or XGA (1024×768) resolution. It is noted that most of the AM-OLEDs are based on the polycrystalline silicon (poly-Si) TFTs technology [83] due to their better electrical performance and operational stability in comparison to the hydrogenated amorphous silicon (a-Si:H) TFTs. However, in 2003, Chi Mei Optoelectronics and IBM Japan [84] showed that the a-Si:H TFTs can in principle be used for AM-OLEDs when a proper and stable a-Si:H TFT process is developed. Since a much advanced a-Si:H TFT technology is already available at low cost from the AM-LCD industry, it would be a great achievement to develop a reliable display technology that takes advantage of the already developed a-Si:H TFT technology.

Apart from the intrinsic technological challenges in developing AM-OLED platform, it has always been considered that the most promising candidates for realizing large-area and flexible organic light-emitting devices are PLEDs, over expensive high-temperature and vacuum-processed OLEDs based on small molecules, because of the printability of polymeric layers in air. Today, despite breakthrough claims being made periodically by equipment makers, small-molecule device production under high-vacuum conditions using shadow masking to create the RGB pixel structure is stuck at "half cut" processing of Gen 4 ($730 \times 920 \, \text{mm}^2$) substrates. The substrate is processed by deposition through multiple masking steps to produce separate RGB pixels in successive operations on half the substrate; the process is then repeated on the other half. Nonetheless, currently, all commercial OLED products are based on small-molecule OLED technology.

PLEDs are mostly based on polyfluorene materials, which represent the device core platform. The device structure incorporating the PLED light-emitting polymer layer has also undergone substantial evolution. The constant struggle for developers is to achieve efficient charge injection, as well as to balance the flux of electrons from the cathode with that of holes from the anode. This is to ensure that electron–hole recombination in the light-emitting polymer (LEP) layer is optimal.

Development of phosphorescent red LEPs has been so successful that development has been halted to allow focus on green and blue emitters. For red light emitters, lifetimes in excess of 200,000 hours (from an initial brightness of $1000\,cd/m^2$) and efficiencies in excess of $20\,cd/A$ have been achieved—a performance that satisfies the requirements for large-screen TVs. The development of a green phosphorescent material with a lifetime in excess of 40,000 hours and an efficiency of $>40\,cd/A$ has also been accomplished. International color standards have been met for both red and green light emitters. Blue devices, having the largest band gap, are the biggest challenge. However, huge advances have been made in the lifetimes and efficiencies of these devices [85].

From what we have highlighted and discussed until now, it is evident that the OLED displays market is slightly more mature than the OLED lighting market. Indeed, two important players in the displays market—LG and Samsung—have recently launched OLED televisions.

The displays market has, for many years, been the focus of attention for OLED development. Now that many of the OLED lifetime and efficiency targets have been achieved for displays, OLED companies can focus on lighting applications. Many technical challenges, however, are common to both applications. For example, there are issues with the yield and reliability of the silicon backplane. In contrast to LCDs, OLEDs are an emissive technology, and this makes them sensitive to defects and power surges in the silicon backplane, causing problems in the display and lighting industries.

The use of defect-tolerant OLEDs, as, for example, proposed by Novaled [86], maintains the appearance of a homogeneously-lit surface—even in the case of electrical short-circuits—owing to the use of a special electrode design. One OLED element consists of two comb-shaped interlocked electrodes, giving a uniform light emission. If one of the stripes short-circuits, the resistance of the stripe will limit the current flow and prevent a further rise in current and temperature, which would otherwise destroy the OLED.

Moreover, it is interesting to note that the OLED lighting technology might have unexpected competitors. While it is unlikely that inorganic thin-film electroluminescence will reach similar parameters in terms of efficiency, brightness, and lifetime, the inorganic LED in combination with light-distributing sheets might be a serious competitor.

Such technologies are currently intensively developed for liquid crystal display backlighting. The current efficiency advantage of the white LED, in combination with the low-cost potential of the light-distribution sheets, might yield a product that is, to the end-user experience, similarly or more attractive than the much more elegant solution offered by the OLED.

2.3   DIODE VERSUS TRANSISTOR ARCHITECTURE FOR LIGHT EMISSION

The excursus on OLED technology highlighted that the development of integrated organic electronic devices is a gateway for a variety of applications and is of great relevance for the general purpose of achieving highly integrated optoelectronic systems.

Particularly, the combination of OLEDs and OTFTs is needed for the development of all-organic active-matrix display technology.

Organic field-effect semiconductor materials are attractive for their homogeneity, low cost, and the variety of means by which they can be deposited, but their best mobilities are similar to that of amorphous silicon.

In the typical TFT architecture, low-mobility field-effect active layers would require a large source–drain voltage to drive the necessary current. This consumes power in the transistor (and sums to the power used to produce light in the OLED), compromising the power savings. In one all-organic active matrix organic light emitting diode (AM-OLED) demonstration, more power was dissipated in the drive transistor than in the OLED it was powering [79]. Mitigating this by increasing the channel width of the drive transistor to the source, more current is not viable given that it would reduce the fraction of pixel area available to the OLED, requiring a higher current density through the electroluminescent emitter to maintain the display brightness, reducing OLED lifetime [87]. Alternatively, the low mobility of the organics could be compensated by making the channel length short, placing the source and drain terminals very close to each other; however, this would increase the expense of high-resolution patterning.

Evidently, other strategies are to be implemented in order to preserve the simplicity of the system architecture and the production costs while enhancing the optoelectronic performances of the photonic components. Indeed, the

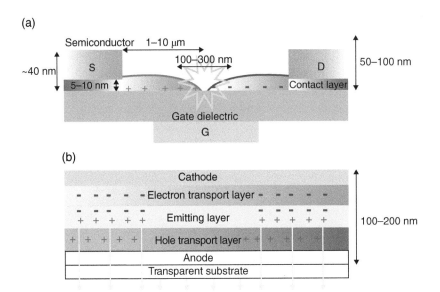

FIGURE 2.10 Schematic representation of the device structures and of the main optoelectronic processes occurring in an OLET and in an OLED. (a) Electron and hole charge transport occurs in-plane at the dielectric/organic interface in OLETs. (b) Charges move vertically across the organic layers in OLEDs. The dimensions of the features of the devices as well as the location of the recombination areas are shown. [*Source*: Muccini [88]. Adapted with permission from Nature Publishing Group.]

combination of different functionalities (i.e., field-effect charge transport and light emission) in a single device would be of great relevance for the further development of organic electronics in integrated components and circuitry.

As we will show throughout this book, OLETs represent a novel class of organic devices capable of combining the current modulating function of a transistor with light emission and could pave the way toward nanoscale light sources and highly integrated organic optoelectronics. In an ideal ambipolar OLET with only one semiconducting layer, the ambipolar regime is characterized by a hole and an electron accumulation layer next to the respective electrodes (namely, source and drain electrode) providing an effective pn junction within the transistor channel (Figure 2.10a). Excitons are thus created by the recombination of in-plane moving electron and hole currents, which are controlled by the gate electrode. Electroluminescence intensity is tuned by the voltages at both the drain (V_{ds}) and the gate (V_g) electrodes.

Due to different bias driving scheme and planar architecture with respect to standard OLEDs, OLETs offer an ideal structure for improving the lifetime and efficiency of organic light-emitting materials.

Hereafter, we will discuss in detail the two concepts on the basis of the working principles of the organic diode- and transistor-like platform for light emission, highlighting the fundamental differences and identifying the advantages of the newly proposed device structure [88].

The main difference between the vertical (OLED) and planar (OLET) device geometry is that charge transport occurs perpendicular to the organic layers (bulk charge transport) and in the plane at the dielectric/organic interface (field-effect charge transport) [89], respectively.

Under the typical biasing conditions, the mobility of electrons and holes can be about four orders of magnitude higher in OLETs than in OLEDs, with direct consequences on material lifetime and exciton emission. It should be noted that in a typical OLED structure, the minority carriers are required to travel only a few tens of nanometers to encounter the carrier of opposite sign and to recombine radiatively. On the contrary, carriers of both signs must travel longer distances (a few hundreds of nanometers to a few micrometers) in a typical ambipolar OLET, which imposes more-stringent requirements on the charge-transport properties of materials. A clear advantage of the OLETs is the potentially higher electroluminescence quantum efficiency inherent to the device structure. In fact, exciton quenching by interaction with charge carriers, with the externally applied electric field and with metal contacts, is drastically reduced (see Figure 2.10).

These aspects are of increasing importance when devices are driven under high-injection conditions for high-brightness emission. In particular, the exciton quenching phenomena in vertical organic optoelectronics devices (such as OLEDs) accounting for the reduction of the overall quantum efficiency in working devices [90] are mainly related to the high electric fields, potentially exceeding 10^8 V/cm [91,92].

The scenario is fundamentally different in the case of organic field-effect transistors, given the planar intralayer charge transport. Indeed, the electric field intensity in the channel of a typical OLET presenting a 500 nm thick dielectric layer remains below 10^6 V/cm for a gate bias up to 100 V. Such a value is usually considered too low to generate sizable exciton dissociation in a single-layer and single-material organic device [10].

Furthermore, relevant differences between the two archetypical structures are noticed also in the processes ruling another well-known source for quenching of internal quantum efficiency, that is, the interaction of excitons with charges. Typically, the excitation energy is transferred via dipole–dipole interaction (Förster transfer) from the excited neutral molecule to a charged molecule and subsequently dissipates nonradiatively. In vertically stacked devices, the charge density is typically about 10^{15} cm^{-3}, which can be correlated to an average charge–charge distance of about 100 nm. Considering that the exciton diffusion length is in the range of tens of nanometers, exciton–charge interaction is negligible in devices with vertical stack geometry [93,94]. On the contrary, the sheet charge density can reach values up to 10^{14} cm^{-2} [95] in organic field-effect transistors (OFETs), which corresponds to a distance of 1 nm among charges, thus well within the diffusion length of excitons [96].

These aspects are of increasing importance when devices are driven under high-injection conditions for achieving high-brightness emission. It is important to highlight that, experimentally, the current density in polymer OLETs, for example, reaches values of 50 A/cm^2 without a decrease in quantum efficiency [97]. Single-crystal devices allow for even higher densities up to 4 kA/cm^2 [98]. The reported values are considerably above those achieved in conventional OLEDs, which are in the order 1–10 A/cm^2 at their point of maximum efficiency.

In addition, the extreme spatial localization of charge carriers in an OLET could be more favorable for an effective spatial separation between the exciton population and the charge carriers. The availability of a third electrode to balance electron and hole currents and therefore to further reduce exciton–charge quenching is the other obvious advantage of OLETs.

In OLEDs, the highly dense electron and hole currents converge to the light-emitting layer, where they form, but spatially coexist with the excitons, and lead to significant exciton–polaron quenching [99–101,102]. Indeed, a seminal theoretical work [103] investigating the impact of various annihilation processes on the laser threshold current density of a multilayer organic laser diode by numerical simulation revealed that the singlet–polaron annihilation is the dominating quenching process. Thus, the focus of OLET development is the possibility to enable new light-emission source technologies and exploit a transport geometry to suppress the deleterious photon losses and exciton quenching mechanisms inherent in the OLED architecture.

The aforementioned inherent outperformance of OLETs with respect to OLEDs in terms of internal quantum efficiency was recently experimentally demonstrated.

Indeed, the first direct experimental comparison was carried out by Capelli *et al.* [104] by fabricating a couple of corresponding OLED and OLET devices presenting the same active region as schematized in Figure 2.11a and b. In the equivalent OLED, the layer sequence, thickness, and film growth parameters are exactly those used for the heterostructure-based OLET fabrication while the electrodes are ITO coated with a poly(3,4-ethylenedioxythiophene) (PEDOT) layer (anode) and Au (cathode).

As the work functions of ITO/PEDOT and Au are similar, the charge-injection conditions in the OLED configuration mimic the OLET case where both the drain and source electrodes are made of gold. The optoelectronic

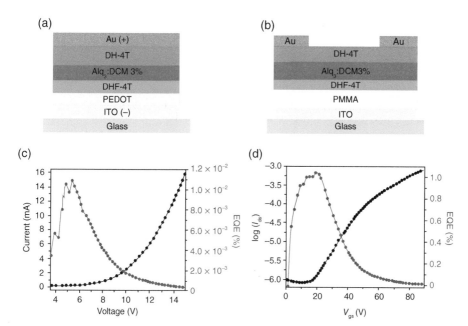

FIGURE 2.11 Schematic structure of the heterostructure-based OLED (a) and OLET (b) with the corresponding optoelectronic characteristics (c and d, respectively). The curve representing the external quantum efficiency is reported in magenta. [*Source*: Capelli *et al.* [103]. Adapted with permission from Nature Publishing Group.]

characteristics of this device (Figure 2.11c) follow a typical *L–I–V* OLED behavior and exhibit a maximum EQE of ~0.012%. This value is almost 100 times lower than the EQE of the corresponding OLET device (Figure 2.11d), which is driven by spanning the bias applied at the gate electrode while keeping the bias applied between the source and drain electrodes constant. This result highlights that the control over the quenching and loss mechanisms is achieved in the planar-geometry field-effect device.

Furthermore, the optimized heterostructure-based OLET (see Chapter 3 for a detailed discussion) was compared with an OLED based on exactly the same emitting layer [105] while the other layers such as HIL, HTL, and ETL were thoroughly optimized. The experimental outcome provided an important figure of merit to fully appreciate the advantages of the heterostructure-based OLET configuration with respect to conventional OLEDs: indeed, a >2*X* factor in the EQE was gained also in this case.

It is interesting to note that the demonstration of the inherent enhancement of the efficiency in OLET has been achieved also in simple single-layer device configuration.

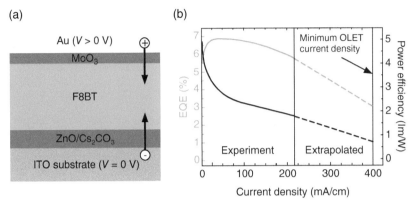

FIGURE 2.12 (a) Illustration of the device structure and emission performance of optimized F8BT OLEDs (with zinc oxide/cesium carbonate and molybdenum trioxide injection layers, as well as 1450 nm thick F8BT). (b) Measured EQE and power efficiency curves dependent on the current densities (data from Reference 106). To compare the values with the F8BT OLETs, the curves are extrapolated further toward higher current densities. [*Source*: Gwinner *et al.* [105], Figure 1, p. 2729. Adapted with permission from Wiley-VCH Verlag GmbH & Co. KGaA.]

Gwinner *et al.* [106] directly compared the light-emission performance at the same current density in the case of OLED and OLET both using F8BT polyfluorene as active layer. With respect to the OLET, the OLED structure based on F8BT had to be optimized by introducing electron- and hole-injection layers formed by ZnO/Cs_2CO_3 and MoO_3, respectively [106] and by increasing F8BT thickness from 60 to 1450 nm in order to achieve the highest EQE (Figure 2.12). Also, in this case, the gold and ITO electrodes are also used in the diode device.

The authors outlined a method for determining the current density value suitable for a reliable comparison of the optoelectronic performances of the two prototypal organic devices. In the case of OLETs, they assessed a lower limit estimation of the current density by dividing the measured current by the product of the thickness of the semiconducting layer and the channel width. In this way, it was demonstrated that OLETs retained high efficiency up to current densities >400 mA/cm², while the highest efficiency F8BT OLEDs made with 1450 nm thick active layers [106] exhibit their maximum efficiency at 50 mA/cm² and show significant efficiency roll-off for higher current densities (Figure 2.12). When the OLED curves are extrapolated using the slope at the largest measured currents, it is extracted that for a minimum OLET current density of $400\,mA/cm^2$, the OLED EQE and power efficiency are expected to have dropped to below 3% and less than 1 lm/W, respectively.

The OLET current density is a lower limit estimate, because the actual current in the OLET is likely to be confined at the proximity of the semiconductor–gate dielectric interface, which would result in even higher (up to 20–60 times) OLET current densities. This provides an important advantage of OLETs over OLEDs, particularly for applications requiring high light intensities such as electrically pumped lasers (see Chapter 6). It is important to note that the comparison between OLETs and OLEDs is meaningful even in terms of other figures of merit apart from EQE, such as luminance emitted per surface area. In the OLET device engineered by *Gwinner et al.*, brightness values up to $8000\,cd/m^2$ can be obtained using the experimentally measured width of the recombination zone for the calculation of the light-emitting area. This value is again comparable to the best values achieved in optimized F8BT OLEDs [107]. As we will discuss in more detail in Chapter 6, the brightness value of the OLET is reduced by typically one order of magnitude when using the effective device area for the indirect calculation, including the nonemitting parts of the channel and electrodes; however, this can be reduced by selecting a small channel length and electrode finger width and using an interdigitated electrode geometry.

However, apart from the benefits of being able to minimize luminescence quenching at metal electrodes and by polarons that we have already mentioned, OLETs exhibit a number of other key advantages over OLEDs. One of the biggest issues for further improving OLEDs is the need to increase the light outcoupling. As we have reported in the previous paragraph, 80% of the emitted light is typically trapped in waveguide modes in the substrate and in the films. In standard bottom-emission device structure, the major cause of the reduced light outcoupling in far-field modes is related in the high refractive index of the transparent electrode [46].

In OLETs, light outcoupling is inherently facilitated, given that the emission region is located far away from the injecting electrodes, with consequent avoidance of metal-induced photon losses. Moreover, photonic structures for light guiding, confinement, and extraction can be easily integrated into the OLET channel underneath the active layer without perturbing the lateral charge transport [108]. This is particularly important for high-resolution displays where the lateral dimensions available for integrating light outcoupling structures are severely limited.

Finally, OLETs also offer an easily-processed device architecture that naturally avoids pinholes and shorts between injection contacts, which are among the main technological problems faced by OLEDs.

The realization of next-generation organic light-emitting devices based on photonic field-effect platform, which outperform current devices in terms of luminance and/or excitation density, is guaranteed once suitable materials capable of combining excellent ambipolar charge-transport properties with high luminescence efficiency are available.

For these reasons, we believe that the application potential of OLETs should not be overlooked and that with further development, OLETs could provide a serious alternative to OLEDs not only for displays but also for lighting applications.

2.4   CONCLUSIONS

In this chapter, the principal characteristics and the working principles of the OLED devices have been briefly reviewed. The high technological impact that OLEDs might have in the broad consumer electronics and lighting markets raises significant interest in OLETs, as this new class of devices might overcome some of the technological hurdle faced by OLEDs when transferred to real manufacturing environments. The structural simplification and the integration of functionalities point to the possibility of developing simplified systems, which might prompt even further the adoption of the organic electroluminescent technologies for light generation. The concept, structure, working principle, and optoelectronic characteristics of OLETs are more thoroughly discussed in Chapter 3.

REFERENCES

[1] M. A. Baldo, D. F. O'Brien, M. E. Thompson, and S. R. Forrest, *Phys. Rev. B* **1999**, *60*, 14422.

[2] S. R. Forrest, *Chem. Rev.* **1997**, *97*, 1793.

[3] T. X. Zhou, T. Ngo, J. J. Brown, M. Shtein, and S. R. Forrest, *Appl. Phys. Lett.* **2005**, *86*, 021107.

[4] R. H. Friend, *et al.*, *Nature (London)* **1999**, *397*, 121.

[5] S. R. Forrest, *Nature (London)* **2004**, *428*, 911.

[6] H. Riel, S. Karg, T. Beierlein, B. Ruhstaller, and W. Riess, *Appl. Phys. Lett.* **2003**, *82*, 466.

[7] H. Kanno, Y. Sun, and S. R. Forrest, *Appl. Phys. Lett.* **2005**, *86*, 263502.

[8] Q. Huang, K. Walzer, M. Pfeiffer, V. Lyssenko, G. F. He, and K. Leo, *Appl. Phys. Lett.* **2006**, *88*, 113515.

[9] V. Bulovic, G. Gu, P. E. Burrows, S. R. Forrest, and M. E. Thompson, *Nature (London)* **1996**, *380*, 29.

[10] M. Pope and C. E. Swenberg, Eds., *Electronic Processes in Organic Crystals* (Oxford University Press, New York, **1999**).

[11] M. Segal, M. A. Baldo, R. J. Holmes, S. R. Forrest, and Z. G. Soos, *Phys. Rev. B* **2003**, *68*, 075211.

[12] T. Tsutsui, C. Adachi, and S. Saito, *Photochemical Processes in Organized Molecular Systems*. Elsevier edn. K. Honda, **1991**, 437.

[13] J. Kido, K. Nagai, and Y. Ohashi, *Chem. Lett.* **1990**, *19*, 657.

[14] J. Kido, H. Hayase, K. Hongawa, K. Nagai, and K. Okuyama, *Appl. Phys. Lett.* **1994**, *65*, 2124.

[15] Z. R. Hong, *et al.*, *Adv. Mater.* **2001**, *13*, 1241.

[16] H. Yersin, *Transition Metal And Rare Earth Compounds III* (Springer-Verlag, Berlin/Heidelberg, **2004**), 241.

[17] Y. Kawamura, J. Brooks, J. J. Brown, H. Sasabe, and C. Adachi, *Phys. Rev. Lett.* **2006**, *96*, 017404.

[18] M. A. Baldo, C. Adachi, and S. R. Forrest, *Phys. Rev. B* **2000**, *62*, 10967.

[19] S. Reineke, K. Walzer, and K. Leo, *Phys. Rev. B* **2007**, *75*, 125328.

[20] B. W. D'Andrade, *et al.*, *Appl. Phys. Lett.* **2001**, *79*, 1045.

[21] T. Iwakuma, *et al.*, *SID Int. Symp. Digest Tech. Papers* **2002**, *33*, 598.

[22] C. Hosokawa, *et al.*, *SID Int. Symp. Digest Tech. Papers* **2004**, *35*, 780.

[23] M. Kawamura, *et al.*, *SID Int. Symp. Digest Tech. Papers* **2010**, *41*, 560.

[24] R. G. Kepler, P. Avakian, J. C. Caris, and E. Abramson, *Phys. Rev. Lett.* **1963**, *10*, 400.

[25] K. Okumoto, H. Kanno, Y. Hamaa, H. Takahashi, and K. Shibata, *Appl. Phys. Lett.* **2006**, *89*, 063504.

[26] D. Y. Kondakov, *J. Appl. Phys.* **2007**, *102*, 114504.

[27] A. Endo, M. Ogasaware, A. Takahashi, D. Yokoyama, Y. Kato, and C. Adachi *Adv. Mater.* **2009**, *21*, 4802.

[28] S. Reineke *Nat. Photonics* **2014**, *8*, 269.

[29] H. Uoyama, K. Goushi, K. Shizu, H. Nomura, and C. Adachi, *Nature (London)* **2012**, *492*, 234.

[30] M. E. Thompson, *MRS Bull.* **2007**, *32*, 694.

[31] M. Segal, M. Singh, K. Rivoire, S. Difley, T. Van Voorhis, and M. A. Baldo, *Nat. Mater.* **2007**, *6*, 374.

[32] R. Meerheim, M. Furno, S. Hofmann, B. Lüssem, and K. Leo, *Appl. Phys. Lett.* **2010**, *97* 253305.

[33] A. Dodabalapur *et al.*, *J. Appl. Phys.* **1996**, *80*, 6954.

[34] M. Thomschke, R. Nitsche, M. Furno, and K. Leo, *Appl. Phys. Lett.* **2009**, *94*, 083303.

[35] S. Reineke, M. Thomschke, B. Lussem, K. Leo *Rev. Mod. Phys.* **2013**, *85*, 1245.

[36] M. Furno, R. Meerheim, S. Hofmann, B. Luessem, and K. Leo, *Phys. Rev. B* **2012**, *85*, 115205.

[37] R. Meerheim, S. Scholz, S. Olthof, G. Schwartz, S. Reineke, K. Walzer, and K. Leo, *J. Appl. Phys.* **2008**, *104*, 014510.

[38] H. Greiner, *Jpn. J. Appl. Phys.* **2007**, *46*, 4125.

[39] B. C. Krummacher, S. Nowy, J. Frischeisen, M. Klein, and W. Brutting, *Org. Electron.* **2009**, *10*, 478.

[40] S. Mladenovski, K. Neyts, D. Pavicic, A. Werner, and C. Rothe, *Opt. Express* **2009**, *17*, 7562.

[41] S. Reineke, G. Schwartz, K. Walzer, and K. Leo, *Phys. Status Solidi RRL* **2009**, *3*, 67.

[42] M. Furno, R. Meerheim, M. Thomschke, S. Hofmann, B. Lussem, and K. Leo, *Proc. SPIE Int. Soc. Opt. Eng.* **2010**, *7617*, 761716.

[43] C. Adachi, M. A. Baldo, M. E. Thompson, and S. R. Forrest, *J. Appl. Phys.* **2001**, *90*, 5048.

[44] G. Gartner, and H. Greiner, *Proc. SPIE Int. Soc. Opt. Eng.* **2008**, *6999*, 69992T.

[45] S. Moller and S. R. Forrest, *J. Appl. Phys.* **2002**, *91*, 3324.

[46] Y. Sun and S. R. Forrest, *Nat. Photonics* **2008**, *2*, 483.

[47] S. Reineke, F. Lindner, G. Schwartz, N. Seidler, K. Walzer, B. Lussem, and K. Leo, *Nature (London)* **2009**, *459*, 234.

[48] Z. B. Wang *et al.*, *Nat. Photonics* **2011**, *5*, 753.

[49] H. Cho, C. Yun, and S. Yoo, *Opt. Exp.* **2010**, *18*, 3404.

[50] J. M. Ziebarth, A. K. Saafir, S. Fan, and M. D. McGehee, *Adv. Funct. Mater.* **2004**, *14*, 451.

[51] K. Ishihara *et al.*, *Appl. Phys. Lett.* **2007**, *90*, 111114.

[52] W. H. Koo *et al.*, *Nat. Photonics* **2010**, *4*, 222.

[53] J. P. Yang, *Appl. Phys. Lett.* **2010**, *97*, 223303.

[54] J. Kido, M. Kimura, and K. Nagai, *Science* **1995**, *267*, 1332.

[55] M. A. Baldo *et al.*, *Nature* **1998**, *395*, 151.

[56] Y.-L. Chang *et al.*, *Adv. Funct. Mater.* **2013**, *23*, 705.

[57] S. O. Jeon, S. E. Jang, H. S. Son, and J. Y. Lee, *Adv. Mater.* **2011**, *23*, 1436.

[58] H. Kanno, R. Holmes, Y. Sun, S. Kena-Cohen, and S. Forrest, *Adv. Mater.* **2006**, *18*, 339.

[59] C. W. Tang and S. A. VanSlyke, *Appl. Phys. Lett.* **1987**, *51*, 913.

[60] S. Lamansky *et al.*, *J. Am. Chem. Soc.* **2001**, *123*, 4304.

[61] M. C. Gather, A. Köhnen, and K. Meerholz, *Adv. Mater.* **2011**, *23*, 233.

[62] L. Xiao *et al.*, *Adv. Mater.* **2011**, *23*, 926.

[63] G. Schwartz, S. Reineke, T. C. Rosenow, K. Walzer, and K. Leo, *Adv. Funct. Mater.* **2009**, *19*, 1319.

[64] J. Y. Tsao *et al.*, T.-Y. Seong, H. Amano, J. Han, and H. Morkoc, Eds., *III-Nitride Based Lighting Emitting Diodes and Applications* (Springer, **2013**, 1B.

[65] H. Sasabe *et al.*, *Adv. Mater.* **2010**, *22*, 5003.

[66] S. Gong *et al.*, *Adv. Funct. Mater.* **2011**, *21*, 6, 1168.

[67] D. H. Kim *et al.*, *Adv. Mater.* **2011**, *23*, 2721.

[68] S. Su, C. Cai, and J. Kido, *Chem. Mater.* **2011**, *23*, 2, 274.

[69] R. Wang *et al.*, *Adv. Mater.* **2011**, *23*, 2823.

[70] S. Su, E. Gonmori, H. Sasabe, and J. Kido, *Adv. Mater.* **2008**, *20*, 4189.

[71] Y. -L. Chang and Z. -H. Lu, *J. Display Tech.* **2013**, *9*, 459.

[72] I. Sokolik, R. Priestley, A. D. Walser, R. Dorsinville, and C. W. Tang, *Appl. Phys.* **1996**, *69*, 4168.

[73] J. Kalinowski, W. Stampor, J. Mezyk, M. Cocchi, D. Virgili, V. Fattori, and P. Di Marco, *Phys. Rev. B* **2002**, *66*, 235321.

[74] T. C. Rosenow, M. Furno, S. Reineke, S. Olthof, B. Lu¨ssem, and K. Leo, *J. Appl. Phys.* **2010**, *108*, 113113.

[75] M. Cocchi, J. Kalinowski, D. Virgili, V. Fattori, S. Develay, and J. A. G. Williams, *Appl. Phys. Lett.* **2007**, *90*, 163508.

[76] Z. L. Zhang, X. Y. Jiang, W. Q. Zhu, B. X. Zhang, and S. H. Xu, *J. Phys. D* **2001**, *34*, 3083.

[77] R. Meerheim, K. Walzer, M. Pfeiffer, and K. Leo, *Appl. Phys. Lett.* **2006**, *89*, 061111.

[78] R. Meerheim, R. Nitsche, and K. Leo, *Appl. Phys. Lett.* **2008**, *93*, 043310.

[79] S. Ohta, T. Chuman, S. Miyaguchi, H. Satoh, T. Tanabe, Y. Okuda, and M. Tsuchida, *Jpn. J. Appl. Phys.* **2005**, *44* Part 1, 3678.

[80] Z. Li, H. Meng, Eds., *Organic Light-Emitting Materials and Devices* (CRC Press Taylor & Francis Group, **2007**).

[81] S. Lee, A. Badano, and J. Kanicki, *Proc. SPIE* **2002**, *4800*, 156.

[82] D. E. Mentley, *Proc. IEEE* **2002**, *90*, 453.

[83] S. Terada, G. Izumi, Y. Sato, M. Takahashi, M. Tada, K. Kawase, K. Shimotoku, H. Tamashiro, N. Ozawa, T. Shibasaki, C. Sato, T. Nakadaira, Y. Iwase, T. Sasaoka, and T. Urabe, *SID Tech. Dig.* **2003**, *34*, 1463.

[84] T. Tsujimura, Y. Kobayashi, K. Murayama, A. Tanaka, M. Morooka, E. Fukumoto, H. Fujimoto, J. Sekine, K. Kanoh, K. Takeda, K. Miwa, M. Asano, N. Ikeda, S. Kohara, S. Ono, C.-T. Chung, R.-M. Chen, J.-W. Chung, C.-W. Huang, H.-R. Guo, C.-C. Yang, C.-C. Hsu, H.-J. Huang, W. Riess, H. Riel, S. Karg, T. Beierlein, D. Gundlach, S. Alvarado, C. Rost, P. Mueller, F. Libsch, M. Mastro, R. Polastre, A. Lien, J. Sanford, and R. Kaufman, *SID Tech. Dig.* **2003**, *34*, 6.

[85] D. Fyfe, *Nat. Photonics* **2009**, *3*, 453.

[86] N. Anscombe, *Nat. Photonics* **2009**, *3*, 458.

[87] T. Tsujioka, H. Fujii, Y. Hamada, and H. Takahashi, *Jpn. J. Appl. Phys., Part 1* **2001**, *40*, 2523.

[88] M. Muccini, *Nat. Materials* **2006**, *5*, 605.

[89] F. Dinelli, M. Murgia, P. Levy, M. Cavallini, and F. Biscarini, *Phys. Rev. Lett.* **2004**, *92*, 116802.

[90] S. Haneder, E. Da Como, J. Fedmann, J. M. Lupton, C. Lennartz, P. Erk, E. Fuchs, O. Molt, I. Munster, and W. G. Schildknecht, *Adv. Mater.* **2008**, *20*, 3325.

[91] W. Stampor, J. Kalinowski, P. Di Marco, and V. Fattori, *Appl. Phys. Lett.* **1997**, *70*, 1935.

[92] J. Kalinowski, J. Mezyk, F. Meinardi, R. Tubino, M. Cocchi, and D. Virgili, *J. Chem. Phys.* **2008**, *128*, 124712.

[93] V. Gulbinas, R. Karpicz, I. Muzikante, and L. Valkunas, *Thin Solid Films* **2010**, *518*, 3299.

[94] N. Pfeffer, D. Neher, M. Remmers, C. Poga, M. Hopmeier, and R. Mahrt, *Chem. Phys.* **1998**, *227*, 167.

[95] G. J. Horowitz, *Mater. Res.* **2004**, *19*, 1946.

[96] W. W. A. Koopman, S. Toffanin, M. Natali, S. Troisi, R. Capelli, V. Biondo, A. Stefani, and M. Muccini, *Nano Lett.* **2014**, *14*, 1695.

[97] J. Zaumseil, C. L. Donley, J.-S. Kim, R. H. Friend, and H. Sirringhaus, *Adv. Mater.* **2006**, *18*, 2708.

[98] T. Takenobu, S. Birsi, T. Takahashi, M. Yahiro, C. Adachi, and Y. Iwasa, *Phys. Rev. Lett.* **2008**, *100*, 15.

[99] H. Yamamoto, T. Oyamada, H. Sasabe, and C. Adachi, *Appl. Phys. Lett.* **2004**, *84*, 1401.

[100] M. A. Baldo, R. J. Holmes, and S. R. Forrest, *Phys. Rev. B* **2002**, *66*, 035321.

[101] E. J. W. List, *et al.*, *Phys. Rev. B* **2001**, *64*, 155204.

[102] J. Staudigel, M. Stöyel, F. Steuber, and J. Simmerer, *J. Appl. Phys.* **1999**, *86*, 3895.

[103] C. Gartner, C. Karnutsch, U. Lemmer, and C. Pflumm, *J. Appl. Phys.* **2007**, *101*, 023107.

[104] R. Capelli, S. Toffanin, G. Generali, H. Usta, A. Facchetti, and M. Muccini, *Nat. Materials* **2010**, *9*, 496.

[105] T. Matsushima and C. Adachi, *Appl. Phys. Lett.* **2006**, *89*, 253506.

[106] M. C. Gwinner, D. Kabra, M. Roberts, T. J. K. Brenner, B. H. Wallikewitz, C. R. McNeill, R. H. Friend, and H. Sirringhaus, *Adv. Mater.* **2012**, *24*, 2728.

[107] D. Kabra, L. P. Lu, M. H. Song, H. J. Snaith, and R. H. Friend, *Adv. Mater.* **2010**, *22*, 3194.

[108] M. C. Gwinner, S. Khodabakhsh, M. H. Song, H. Schweizer, H. Giessen, and H. Sirringhaus, *Adv. Funct. Mater.* **2009**, *19*, 1360.

3

ORGANIC LIGHT-EMITTING TRANSISTORS: CONCEPT, STRUCTURE, AND OPTOELECTRONIC CHARACTERISTICS

In recent years, intense multidisciplinary research in molecular design and engineering, materials science, and device technology led to significant progress in understanding and controlling the fundamental properties of materials, which enabled the fabrication of performing organic devices. The best-performing organic field-effect transistors (OFETs) described so far are based on a limited number of low-molecular-weight or polymeric materials such as pentacene, oligothiophenes, poly(phenylene vinylene), poly(thiophene), or poly(fluorene) derivatives. Despite the large variability in chemical structures, all of these materials have in common the unidimensionality of their elemental unit, which results in anisotropic charge transport and optical properties. An important consequence of this anisotropy is that the realization of efficient electronic or photonic devices requires a precise control of the material organization on the substrate. In addition to the tight molecular packing and strong intermolecular interactions needed to reach high charge-carrier mobility, proper control of the orientation of the conjugated chains on the substrate is imperative to obtain efficient charge transport in the desired direction. It has been shown that vertical orientation of π-conjugated molecules on a substrate results in a maximum charge-carrier mobility in the direction of π-stacking and, hence, across the channel of a lateral field-effect device [1–3]. While conjugated polymers could be *a priori* viewed as isotropic materials, it has been shown that the charge-transport process in thin films has

Organic Light-Emitting Transistors: Towards the Next Generation Display Technology, First Edition. Michele Muccini and Stefano Toffanin.
© 2016 John Wiley & Sons, Inc. Published 2016 by John Wiley & Sons, Inc.

a strong 2D character, with the highest in-plane mobility in the crystallized zones of the polymer when chains are oriented edge-on on the substrate [4]. On this basis, it has been shown that increasing the order in the polymer by the use of high-boiling point solvents for film casting or by thermal treatment led to a dramatic improvement of the performances of OFETs [5]. These observations implicitly point to single crystals as ideal model systems toward which the optimization of the experimental conditions of thermal evaporation, solution process, and/or postprocessing treatment of thin films aims at. The final target is a thin-film structure where the packing arrangement of the π-conjugated chains in the material would resemble as much as possible that of a single crystal. Using single crystals makes it easier to analyze the effects of defects and impurities on the device characteristics and to get closer to the limits of the intrinsic electrical and optical properties [6]. However, thin films have a much higher technological potential, even though their morphological and electrical properties are strongly affected by the processing conditions and by the characteristics of the substrate on which they are deposited. Indeed, the insulator layer influences both the growth process and the electrostatic potential in the first layers of the semiconductor, hence determining the device behavior [7].

In this chapter, the fundamental working principles of organic light-emitting transistors (OLETs) are discussed and the key features of the most common device architectures used to generate light from a field-effect transistor (FET) structure are reviewed. The electrical and optoelectronic characteristics of OLETs based on organic thin films, multilayer structures, as well as single crystals are analyzed to highlight the distinguishing features originated by the specific active-layer components and structure.

3.1 WORKING PRINCIPLES OF OLETs

One of the most relevant characteristics of organic semiconductors in view of their use as active materials in OLETs is that they are relatively wide-band-gap semiconductors with band gaps in the optical range (2–3 eV). Indeed, this property makes them attractive as emissive semiconductors. By employing state-of-the-art methods for their synthesis and purification, it is possible to control the concentration of impurities left over from the synthesis to parts per million levels [8,9]. Therefore, even unintentional extrinsic doping levels are low enough, and these materials can be considered, for any practical effects, intrinsic semiconductors. When incorporating these intrinsic organic semiconductors into FET configurations—most commonly thin-film transistors (TFTs)—to evaluate their charge-transport characteristics in combination with a specific gate dielectric, for example, the prototypical SiO_2, many of these materials exhibit hole accumulation behavior for negative applied gate voltages. However, when the gate voltage polarity is reversed to positive values, the formation of an electron accumulation layer is much less commonly observed. For many organic semiconductor-based FETs, only p-channel operation seems possible. For this reason, such materials have been called "p-type" organic semiconductors. The realization of "n-type" organic FETs usually involved the synthesis of special organic semiconductors with high electron affinities, comprising specific electron withdrawing groups. These were then called "n-type" organic semiconductors.

In recent years, it has become clear that the chemical structure of the organic semiconductor is not the only factor that determines whether an organic FET exhibits predominantly p-channel or n-channel behavior. Processing and characterization conditions, device architecture, and choice of electrodes are important as well. It is thus not appropriate to speak of p-type or n-type materials, but one should rather refer to p-channel or n-channel transistors. A key discovery was the identification of the crucial role of the gate dielectric and the identification of electron trapping mechanisms in devices based on SiO_2 gate dielectrics. This subsequently led to the general observation of n-channel and ambipolar characteristics in a broad range of polymer semiconductor FETs based on trap-free gate dielectrics [10]. This evidence together with other recent experimental and theoretical studies suggested that organic semiconductors are intrinsically ambipolar and thus capable of conducting both electrons and holes in suitable device configurations and under inert testing conditions.

Here, we give an overview of the working principles of organic FETs and describe how both hole transport and electron transport take place in these devices, which can be used to fabricate ambipolar and light-emitting transistors. Particularly, we introduce organic FET current–voltage characteristics, discuss how to extract physical information from them, and highlight the parameters affecting the device performance.

Light-emitting transistors take advantage of the efficient radiative recombination of holes and electrons in organic semiconductors, which leads to light emission from the transistor channel under certain biasing conditions, and, from a more fundamental point of view, offer the possibility of the direct visualization of ambipolar transport and recombination characteristics.

The active part of the device is constituted of an organic semiconductor thin film equipped with two electrodes: the source and the drain. The distance between the source and the drain is called the channel length L while the transverse dimension of the structure is the channel width W. A third electrode, the gate, is laid out along the channel between the source and the

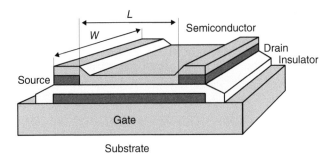

FIGURE 3.1   Three-dimensional view of an organic thin-film transistor.

drain; this electrode is electrically isolated from the semiconductor film by a thin insulating film, hence forming a metal–insulator–semiconductor (MIS) structure (see Figure 3.1).

A unique feature of FETs is that, unlike diodes, these devices are essentially bidimensional. That is, they are governed by two independent perpendicularly oriented electric fields, originated by the potential difference between the source and gate electrodes (V_{gs}) and by the potential difference between the drain and source electrodes (V_{ds}). Usually, the source electrode is grounded (see Figure 3.2).

In unipolar devices, where only one type of charge carrier is transported, the source is the charge-injecting electrode since it is more negative than the gate electrode when a positive V_{gs} is applied (electrons are injected) and more positive than the gate electrode when a negative voltage is applied (holes are injected). In ambipolar devices, where both types of charge carriers can be transported, both the source and drain electrodes can act simultaneously as injecting and collecting electrodes, under suitable biasing conditions as will be discussed later.

Figure 3.2b–d illustrates the basic operating regimes and associated current–voltage characteristics of an n-transport OFET. Given that the organic semiconductor is an n-type material, a positive gate voltage is applied in order to induce negative charges (electrons) at the insulator/semiconductor interface that are injected from the grounded electrode. In the case of a p-type organic semiconductor, a negative V_{gs} is applied in order to accumulate positive charges (holes).

The number of accumulated charges is proportional to V_{gs} and the capacitance C_i of the insulator.

However, not all induced charges are mobile and will thus contribute to the current in a FET. Deep traps first have to be filled before the additionally induced charges can be mobile. That is, a gate voltage has to be applied that is higher than a threshold voltage V_t, and thus, the effective gate

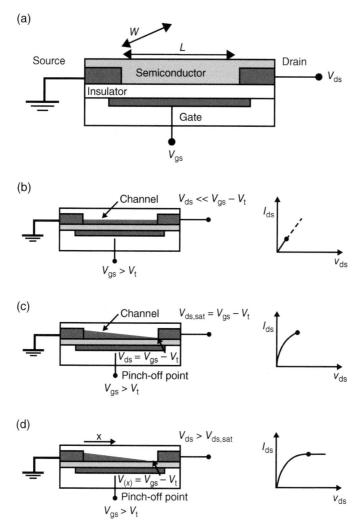

FIGURE 3.2 (a) Schematic structure of a field-effect transistor and applied voltages: L, channel length; W, channel width; V_{ds}, drain voltage; V_g, gate voltage; V_t, threshold voltage; and I_{ds}, drain current. (b–d) Illustrations of operating regimes of field-effect transistors: (b) linear regime; (c) start of saturation regime at pinch-off; (d) saturation regime and corresponding current–voltage characteristics.

voltage is $V_{gs} - V_t$. On the other hand, donor (for n-type) or acceptor (for p-type) states and interface dipoles can create an internal potential at the interface. Thus cause accumulation of charges in the channel when $V_{gs} = 0$ so that, in some cases, an opposite voltage has to be applied to turn the channel off [11,12].

When no source–drain bias is applied, the charge-carrier concentration in the transistor channel is uniform. A linear gradient of charge density from the carrier-injecting source to the extracting drain forms when a small source–drain voltage is applied ($V_{ds} \ll V_{gs}$, Figure 3.2b). This is the linear regime, in which the current flowing through the channel is directly proportional to V_{ds}. The potential $V(x)$ within the channel increases linearly from the source ($x=0$, $V(x)=0$) to V_{ds} at the drain electrode ($x=L$, $V(x)=V_{ds}$).

When the source–drain voltage is further increased, a voltage potential value $V_{ds}=V_{gs}-V_t$ is reached, at which the channel is "pinched off" (Figure 3.2c). That means a depletion region forms next to the drain because the difference between the local potential $V(x)$ and the gate voltage is now below the threshold voltage. A space-charge-limited saturation current $I_{ds,sat}$ can flow across this narrow depletion zone as carriers are swept from the pinch-off point to the drain by the comparatively high electric field in the depletion region.

Further increasing the source–drain voltage will not substantially increase the current but leads to an expansion of the depletion region and thus a slight shortening of the channel. Since the potential at the pinch-off point remains $V_{gs}-V_t$ and thus the potential drop between that point and the source electrode maintains constant, the current saturates at a level $I_{ds,sat}$ (Figure 3.2d).

Typically, transistors with short channel lengths require that the gate dielectric thickness is one order of magnitude thinner than the channel length [13] in order to ensure that the field created by the gate voltage determines the charge distribution within the channel and is not dominated by the lateral field due to the source–drain voltage. Otherwise, a space-charge-limited bulk current will prevent saturation and the gate voltage will not determine the "on" or "off" state of the transistor [14,15].

The current–voltage characteristics in the different operating regimes of FETs can be described analytically by assuming that (i) the transverse electric field induced by the gate voltage is much higher than the longitudinal field induced by the gate bias (gradual channel approximation) and (ii) the mobility is constant all over the channel.

Assumption (i) is justified by the geometry of the device since the distance from the source to the drain is often much larger than the thickness of the insulator. Assumption (ii) is almost always fulfilled in inorganic semiconductors, while is only true at a first approximation in real organic semiconductors.

At a given gate potential, higher than the threshold voltage V_t, the induced mobile charges Q per unit area at the source contact are related to V_{gs} via

$$Q = C_i \left(V_{gs} - V_t \right) \tag{3.1}$$

where C_i is the capacitance per unit area of the gate dielectric.

In Equation 3.1, the channel potential is assumed to be zero. However, the induced charge density depends on the position along the channel (x), which is considered in the following equation:

$$Q = C_i \left(V_{gs} - V_t - V(x) \right) \tag{3.2}$$

If charge diffusion is neglected, the source–drain current (I_{ds}) induced by the carriers is

$$I_{ds} = W \mu Q E(x) \tag{3.3}$$

where W is the channel width, μ is the charge mobility, and $E(x)$ is the electric field at the x position.

By considering $E(x) = dV/dx$ and substituting Equation 3.2 into Equation 3.3, we find

$$I_{ds}\, dx = W \mu C_i \left(V_{gs} - V_t - V(x) \right) dV \tag{3.4}$$

The gradual channel expression for the drain current can then be obtained by integration of the current increment from $x=0$ to L, that is, from $V(x)=0$ to V_{ds}, by assuming that the mobility is independent of the carrier density and hence the gate voltage:

$$I_{ds} = \left(\frac{W}{L} \right) W \mu C_i \left[\left(V_{gs} - V_t \right) V_{ds} - \left(\frac{1}{2} \right) V_{ds}^2 \right] \tag{3.5}$$

In the linear regime with $V_{ds} \ll V_{gs}$, Equation 3.5 can be simplified to

$$I_{ds} = \left(\frac{W}{L} \right) W \mu C_i \left[\left(V_{gs} - V_t \right) V_{ds} \right] \tag{3.6}$$

The drain current is directly proportional to V_{gs}, and the field-effect mobility in the linear regime can thus be extracted from the gradient of I_{ds} versus V_{gs} at constant V_{ds} (also applicable for gate-voltage-dependent mobilities).

As described earlier, the channel is pinched off when $V_{ds} = V_{gs} - V_t$. The current cannot increase substantially anymore and saturates $(I_{ds,sat})$. Thus, Equation 3.5 is no longer valid.

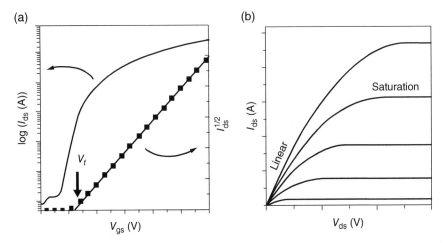

FIGURE 3.3 Representative current–voltage characteristics of an n-channel organic field-effect transistor: (a) transfer characteristics in the saturation regime ($V_{ds} > V_{gs} - V_t$), indicating the threshold voltage V_t, where the linear fit to the square root of the drain current intersects with the x-axis; (b) output characteristics indicating the linear and saturation regimes.

Neglecting channel shortening due to the depletion region at the drain, the saturation current can be obtained by substituting V_{ds} with $V_{gs} - V_t$, yielding

$$I_{ds,\ sat} = \left(\frac{W}{2L}\right)\mu C_i \left(V_{gs} - V_t\right)^2 \tag{3.7}$$

The field-effect mobility in the linear regime (μ_{lin}) can be extracted from the gradient of I_{ds} versus V_{gs} at constant V_{ds}.

$$\mu_{lin} = \frac{\partial I_{ds}}{\partial V_{gs}} \frac{L}{W\, C_i\, V_{ds}} \tag{3.8}$$

The I–V characteristic obtained by sweeping V_{gs} at constant V_{ds} is called *transfer curve* while the characteristic obtained by sweeping V_{ds} at constant V_{gs} is called *output curve* (Figure 3.3a and b).

In the transfer curve in the saturation regime, the square root of the drain current should be linearly dependent on the gate voltage, and its gradient is proportional to the mobility according to Equation 3.7. Extrapolating the linear fit to zero yields the threshold voltage V_t (Figure 3.3a).

Figure 3.3b shows typical output characteristics of a polymer n-channel transistor with a channel length of 200 µm. From the output characteristics, the linear regime at low V_{ds} and the saturation regime at high V_{ds} are evident.

μ_{lin} can be calculated for $V_{gs} - V_t \gg V_{ds}$ by the following two derivatives

$$
g_m = \left.\frac{\partial I_{ds}}{\partial V_{gs}}\right|_{V_{ds}} = \frac{W}{L} C_i \mu_{lin} V_{ds}
$$

$$
g_d = \left.\frac{\partial I_{ds}}{\partial V_{ds}}\right|_{V_{gs}} \approx \frac{W}{L} C_i \mu_{lin} \left(V_{gs} - V_t\right)
$$

(3.9)

Mobility can be calculated from the variation of g_m with V_{ds} or g_d with V_{gs}

$$
\frac{dg_m}{dV_{ds}} = \frac{dg_d}{dV_{gs}} = \frac{W}{L} C_i \mu_{lin}
$$

(3.10)

dI_{ds}/dV_{gs} (and so the mobility) can be considered the angular coefficient of the linear interpolation of the transfer curve at high V_{gs} (linear regime).

The device mobility can also be calculated in the saturation region from the slope of the linear part of $(I_{ds,sat})^{1/2}$ versus V_{gs} plot, called *locus curve*. A locus curve is a plot obtained by collecting I_{ds} current during the sweeping of the drain and gate voltages, which are kept equal. Within this condition, the pinch-off is always reached near the drain contact. If $(I_{ds,sat})^{1/2}$ versus V_{gs} or V_{ds} is plotted, the mobility and the threshold voltage values can be extrapolated from the linear interpolation of the curve

$$
\sqrt{I_{ds}^{sat}} = \sqrt{\frac{W}{2L} \mu_{sat} C_i} \left(V_{gs} - V_t\right)
$$

(3.11)

as angular coefficient and V_{gs}-axis intercept, respectively.

Mobility values calculated in the linear and in the saturation regimes are often different, the saturation one being usually higher. This happens because the conduction channel resistance in saturation is higher than in the linear case; hence, contact resistance is less critical than in the linear region. However, the two mobility values must be equal in devices with good injection contacts.

In conventional metal-oxide-semiconductor field-effect transistors (MOSFETs), the threshold voltage is an important technological parameter that plays a crucial role in circuit modeling. A review of different methods for extracting the threshold voltage can be found in Ortiz-Conde *et al.* [16].

Threshold voltages can originate from several effects and depend strongly on the semiconductor and dielectric used. Built-in dipoles, impurities, interface states, and, in particular, charge traps contribute to the threshold voltage [17]. Note that, independent of the cause of V_t, it can be reduced by increasing the gate capacitance and thus inducing more charges at lower applied voltages.

Given several effects, such as gate-voltage-dependent mobility and contact resistance, the actual curve represented by Equation 3.11 is not a real straight line. Instead, it presents an upward curvature at low gate voltages and a downward curvature at high voltages. On this kind of curve, the extracted mobility and threshold voltage largely depend on the voltage range used for the linear fitting. The most common approach is to center the linear regression curve at the inflection point; however, it is important noting that this approach is purely empirical.

Another important parameter to analyze the OFET performances is the *ON/OFF ratio*. It is the ratio of the drain–source current flowing when the gate bias equals the drain one at the maximum voltage to the one measured at the same drain potential but with the gate bias equal to 0 [18]. This value should be as large as possible to obtain a clear switching behavior of the transistor. If the contact resistance can be neglected, the ON current mainly depends on the organic semiconductor mobility and on the capacitance of the insulator used as dielectric. The OFF current is determined by the organic semiconductor bulk conduction.

In an ambipolar transistor, both holes and electrons are simultaneously injected in the channel by the source and drain electrodes, when they are biased at a suitable voltage. This voltage has to be higher in absolute value with respect to the hole and electron gate threshold voltages. In the case of ideal ambipolar transistor with only one semiconducting layer, the ambipolar regime is characterized by a hole and an electron accumulation layer that meet at some point within the transistor channel. From the mathematical point of view, the model that describes an intrinsic ambipolar transistor is based on the coupling of two *noninteracting* unipolar p-type and n-type OFETs. The coupling between the two OFETs is defined by the edge conditions of the three contact voltages.

The position of the electroluminescence emission region is moved within the channel according to the applied gate and source–drain voltages and the charge-carrier mobility ratio.

Let us assume a transistor at a given positive drain voltage V_{ds} and start with a positive gate voltage of $V_{gs} = V_{ds}$. Just as in a unipolar transistor, the gate is more positive than the source electrode, and thus, electrons are injected from the source into the accumulation layer and drift toward the drain, given that $V_{ds} > V_{t,e}$ with $V_{t,e}$ the threshold for electron accumulation. Only one polarity of charge carriers is present, and ambipolar OFET is said to be working in unipolar region.

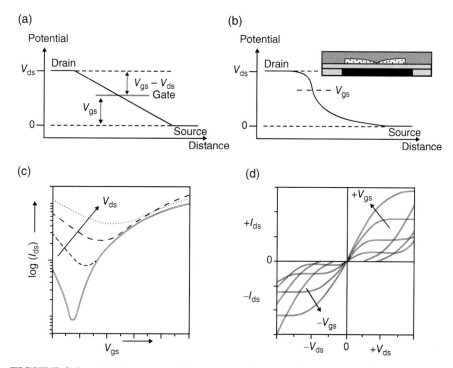

FIGURE 3.4   (a) Illustration of the source, drain, and gate potentials with respect to each other in a field-effect transistor. (b) Channel potential in a field-effect transistor in the ambipolar regime with two separate channels of holes and electrons that meet within the transistor channel, where opposite charge carriers recombine (inset). (c) Calculated transfer characteristics for an ambipolar transistor with equal hole and electron mobilities and slightly different threshold voltages in a semilog plot for positive gate voltages and different positive source–drain voltages. (d) Calculated ambipolar output characteristics for the same transistor for positive (first quadrant) and for negative (third quadrant) V_g and V_{ds}, respectively.

When V_{gs} is smaller than V_{ds}, the gate potential is more negative than that of the drain electrode by $V_{gs} - V_{ds}$. While, the source will not inject electrons given $V_{gs} < V_{t,e}$, the drain electrode in an ambipolar transistor will inject holes into the channel if $V_{gs} - V_{ds} < V_{t,h}$, where $V_{t,h}$ is the threshold for hole accumulation. Thus, a hole current will flow and the measured I_{ds} current is increased differently from the case of unipolar n-channel transistor that would be now in an off state. If the two conditions on V_{gs} are simultaneously fulfilled, such as $V_{gs} > V_{t,e}$ and $V_{gs} - V_{ds} < V_{t,h}$ (Figure 3.4a), both electrons and holes are present in the channel. This regime is called the ambipolar regime, in contrast to the unipolar regime, where only charges with only one polarity are present in the channel for any particular biasing conditions.

In a typical ambipolar OFET, the electrical curves show characteristic curves not displayed by unipolar devices.

In the *I–V* transfer curves in the saturation regime, instead the behavior of an ideal ambipolar OFET is represented by a typical *V-shaped* curve. Each branch of the curve represents a different charge carrier as a function of the polarity of the gate contact. For gate voltage values in the proximity of the edges of the curve, there is a unipolar I_{ds} current whose charge carrier type is defined by the local difference of applied voltages at the source, drain, and gate electrodes. The drain–source current minimum (so that, the bottom of the V-shaped curve) is reached when the two charge-carriers currents are balanced (maximum of ambipolar behavior of the OFET) (Figure 3.4c). Since both charge carriers can flow through the channel, the I_{ds} output current does not completely saturate, but it is composed of two phases. In the first phase, when one charge carrier is flowing, the behavior is similar to that of a unipolar OFET, and eventually, the current can reach a temporary plateau. When the gate voltage reaches a local bias that permits the injection of the opposite charge carrier, the I_{ds} output current begins to raise quadratically (Figure 3.4d).

Within the graduate channel approximation, it is possible to derive a simple analytical expression for the ambipolar regime [19]. If we assume an infinite recombination rate of holes and electrons, all injected holes and electrons have to combine and thus the source–drain current equals the electron and hole current for each channel. Combining the expressions for the saturated currents for the hole and electron channels, the source–drain current is given by

$$\left| I_{ds} \right| = \frac{W C_i}{2L} \left\{ \mu_e \left(V_{gs} - V_{t,e} \right)^2 + \mu_h \left(V_{ds} - \left(V_{gs} - V_{t,h} \right) \right)^2 \right\} \qquad (3.12)$$

The position x_0 of the recombination zone within the channel is obtained by

$$x_0 = \frac{L \left(V_{gs} - V_{t,e} \right)^2}{\left(V_{gs} - V_{t,e} \right)^2 + \dfrac{\mu_h}{\mu_e} \left(V_{ds} - \left(V_{gs} - V_{t,h} \right) \right)^2} \qquad (3.13)$$

Equation 3.13 confirms that the position of the recombination zone, and therefore of the light-emitting area, depends on the applied voltages and on the ratio of the hole- and electron-mobility values.

Building on these principles, it will be possible to analyze the electrical and optoelectronic characteristics of the OLET devices based on single-layer thin films, multilayer structures, and single crystals that are reported in the next section.

3.2 DEVICE STRUCTURES

As we have already briefly discussed, a field-effect organic (light-emitting) transistor is composed by a semiconducting organic material (or by a composition of semiconducting organic materials) working as a *channel*, where current flows. At one extremity of the channel, there is an electrode called *source*, and at the opposite side, there is a second electrode called *drain*. As we have already shown, the physical dimensions of the channel are fixed but the portion actually used in the field-effect conduction is varied by applying a voltage to the third electrode called *gate*. The field-effect conductivity of the transistor depends on the portion of the channel open to the current.

The channel is in contact with a dielectric layer working as a capacitor and allows current modulation through the gate voltage (see Section 3.1).

There are two main architectures that can be chosen for OFET fabrication: the *bottom contact* and the *top contact* configurations, which differ in the sequential order of the key steps in the fabrication process flow. The source/drain contacts are deposited either before or after the semiconductor layer is deposited to create a bottom contact or top contact device, respectively. One can also build the transistor on top of the semiconductor layer (the so-called *top gate* architectures), in which the insulator and gate contact are sequentially deposited on top of either of the two source–drain contact configurations. All four of these OFET architectures are shown schematically in Figure 3.5.

In the bottom gate–bottom contact (BG–BC) configuration, the drain and the source electrodes are positioned directly on the dielectric film. The gate contact is under the dielectric; it usually works also as a substrate, and the active material is grown on top of the dielectric. Differently from the bottom gate–top contact (BG–TC) case, in this configuration, three interfaces operate in the same region: organic semiconductor/contacts, contacts/dielectric, and organic semiconductor/dielectric. The semiconductor growth at the complex triple interface is not fully understood, although it is clear that the bottom contact configuration typically creates greater contact resistance than in the case of top contacts.

Furthermore, treating the dielectric surface to improve the OFET performances for the BG–BC is difficult, if not impossible, because the presence of the electrodes prevents a perfect treatment of the dielectric edges.

However, such devices are easier to process (typically for the classic SiO_2 dielectric, the drain and the source contacts are patterned by photolithography), as they involve less fabrication steps with respect to the BG–TC, and the possible damage of the active layer due to the deposition of the source and drain electrodes is ruled out.

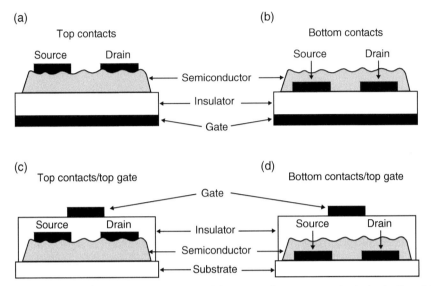

FIGURE 3.5 Four possible OFET architectures (in cross section), including (a) top contacts, (b) bottom contacts, (c) top contacts with a top gate, and (d) bottom contacts with a top gate.

In addition, very small channel dimensions (W, $L < 10\,\mu m$) can be prepatterned on the insulator using conventional photolithography. A limitation to the bottom contact configuration, however, is that film morphology in the vicinity of the contacts is often nonideal. A number of researchers have demonstrated that the organic semiconductor grain sizes are very small near the contacts, presumably due to heterogeneous nucleation phenomena [20]. Pentacene molecules, for example, tend to "stand up" with the long axis of the molecule perpendicular to the plane of the substrate when deposited on the commonly used insulator SiO_2 [21]. When deposited on top of gold contacts, however, strong interactions between the pentacene π-clouds and the metal surface lead to tiny grains at the contact and, in some cases, voids are observed [20].

In the BG–TC configuration, the drain and source electrodes are deposited on top of the organic semiconductor. Underneath, there are the dielectric layer and the gate electrode that can work also as substrate. In this geometry (as in the top gate–bottom contact case), the injection zone (contacts/organic semiconductor) is well separated from the conducting zone (organic semiconductor/dielectric). Moreover, in this device configuration, the chemical functionalization of the dielectric surface required to control the active-layer growth is a simple and effective process.

Top contact OFETs (Figure 3.5a) generally exhibit the lowest contact resistances. This is likely because of the increased metal–semiconductor contact area in this configuration. A major contribution to contact resistance in the top contact configuration is what is called *access resistance*. Access resistance results from the requirement that charge carriers must travel from the source contact on top of the film down to the accumulation layer (the channel) at the semiconductor–insulator interface and then back up to the drain contact to be extracted.

In order to minimize access resistance, the thickness of the organic semiconductor layer should not be too large. However, some researchers have proposed that access resistance is less than might be expected for top contact OFETs because the contact metal penetrates the film down to the accumulation layer (perhaps due to large peak-to-valley roughness of the semiconductor film or the nature of the metal deposition process) [22]. With the bottom contact architecture (Figure 3.5b), access resistance is not an issue because the contacts are in the same plane as the OFET channel.

Of the two top gate OFET architectures (Figure 3.5c and 3.5d), the top contact/top gate configuration (Figure 3.5c) is the more favorable of the two because bottom contact/top gate devices suffer from access resistance. However, it should be noted that both top gate architectures face the additional concerns of semiconductor top surface roughness (since this is where the channel will form) and forming a stable interface between the insulator and the top of the semiconductor film. Solution deposition of the top insulator material, for example, may damage the underlying semiconductor film. Finally, regarding the alignment of the top gate contact to the OFET channel in top gate devices, care must also be taken to ensure that the gate reaches completely across the entire length of the device. If the length of the gate electrode is less than the channel length or if the gate is simply misaligned, additional contact resistance will be introduced as a result of ungated semiconductor regions at one or both of the contacts. This device configuration has been fully exploited in the fabrication of state-of-the-art single-layer polymeric OLETs, implementing polyfluorene-derivative polymers (e.g., F8BT, F8TBT) as active materials and polymethylmethacrylate (PMMA) layer as dielectric [23].

In general, in ambipolar OLETs, the "two-contact" device geometry was implemented in ambipolar OLETs, that is, high-work-function metal for the electron-injection electrode and low-work-function metal for the hole-injection electrode, in order to balance the electron and hole injection [24,25]. From a fabrication standpoint, depositing two different contact materials at either end of the transistor channel is a further technological step in the device realization, which, however, guarantees a lower contact resistance and a higher device efficiency.

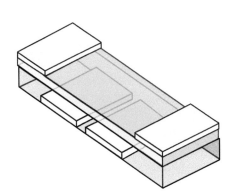 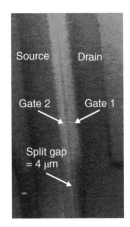

FIGURE 3.6 A schematic of the architecture of a split-gate OLET, in bottom-gate configuration. A picture of SG-LEFET shows the split gate; the width of the gap in the split gate is 4 μm. [*Source*: Hsu *et al.* [26]. Adapted with permission from Wiley.]

In the search of higher device efficiency and higher control of the device functionality, other OLET device architectures with specifically tailored features have been proposed.

Hsu *et al.* [26] realized an OLET endowed with a split gate, in order to overcome the material constraint and simultaneously improve the brightness and external quantum efficiency (EQE) of the device. A split-gate OLET is a four-terminal device with Gate 1, Gate 2, source, and drain electrodes (Figure 3.6). Based on the operating voltages, (i.e., the bias polarities on Gate 1 and Gate 2), a split-gate FET can be operated as a unipolar transistor, a bipolar transistor, or a diode in different I–V quadrants [27]. Because of the independent control of injection at the two distinct gate electrodes, transport of specific carrier species can be enhanced or switched off. By using the split-gate OLET, it is demonstrated that the recombination zone can be tuned at the center of the transistor channel with maximum brightness of 609 cd/m^2 at 1.27% EQE.

In the first split-gate OLET device, the active material was the widely used F8BT polymer [24]. In the reported case, the brightness value achieved with this complex device structure was much less than that obtained by other groups when optimizing the standard single- and top-gate configuration (see Chapter 4) [28].

An alternative device structure, which combines lateral field-effect charge transport and vertical OLED-like recombination, has been reported by Schols *et al.* [29] (Figure 3.7).

The device is a hybrid between a diode and a FET. Compared to conventional OLEDs, the metallic cathode is displaced by one to several micrometers

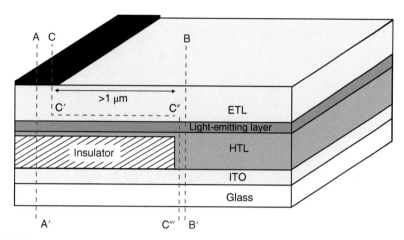

FIGURE 3.7 Schematic of the device structure proposed by Schols *et al.* The device is a hybrid between an OLED and a field-effect structure and comprises an organic hole-transport layer, an organic electron-transport layer, and an organic light-emitting layer. The dashed lines AA', BB', and CC'C"C indicate three different cross sections. [*Source*: Schols *et al.* [29]. Adapted with permission from Wiley.]

from the light-emitting zone. This micrometer-sized distance can be bridged by electrons with enhanced field-effect mobility. The absence of a metallic cathode covering the light-emission zone permits top emission and could reduce optical absorption losses in waveguide structures.

The device realized by Schols *et al.* was fabricated using poly(triarylamine) (PTAA) as the hole-transport material, tris(8-hydroxyquinoline) aluminum (Alq_3) doped with 4-(dicyanomethylene)-2-methyl-6-(julolindin-4-yl-vinyl)-4H-pyran (DCM_2) as the active light-emitting layer, and N,N'-ditridecylperylene-3,4,9,10-tetracarboxylic diimide (PTCDI-$C_{13}H_{27}$), as the electron-transport material. The obtained external quantum efficiencies are as high as for conventional OLEDs comprising the same materials.

3.3 THIN-FILM OLETs

3.3.1 Single-Layer OLETs

Light emission from an organic FET was first detected in a single-layer unipolar tetracene FET [30]. The tetracene active layer was deposited onto a Si/SiO$_2$ substrate with an interdigitated pattern of gold source and drain electrodes in BG–BC configuration.

(a)

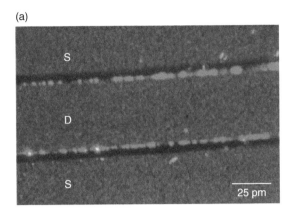

(b)

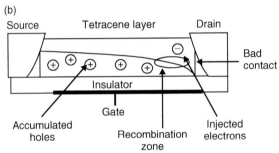

FIGURE 3.8 (a) Picture of the first operating OLET based on tetracene taken through an optical microscope. (b) Schematic of the device architecture with the possible representation of the light formation mechanism: in hole accumulation mode, electrons can be injected into tetracene far away from the gate dielectric interface given the uneven contact at the drain electrode. [*Source*: Hepp *et al.* [30], Figure 7. Adapted with permission of American Physical Society.]

Electroluminescence was unexpectedly observed from a p-channel FET since this implies that both holes and electrons are present in the channel to form excitons and recombine radiatively.

Optical imaging (Figure 3.8a) revealed that light emission took place exclusively at the edge of the drain electrodes and not within the channel. In order to explain empirically the possible electron injection at the drain electrode, Hepp *et al.* [30] hypothesized different charge-injection mechanisms at the source and drain electrodes due to imperfections in device fabrication. In particular, poor electrical contact due to electrode underetching was deemed responsible for the presence of intense electric fields in the vicinity of the drain electrode (Figure 3.8b), which in turn prompted electron injection into the organic layer [31].

Santato *et al.* [32] found later that the electrode undercut was not necessary to achieve light emission in tetracene-based devices with gold electrodes. They proposed a phenomenological model in which a steep voltage drop at the electrode/organic interface induces a distortion in the highest occupied molecular orbital and lowest unoccupied molecular orbital (HOMO and LUMO, respectively) of tetracene molecules near the metal electrode, thereby determining the conditions for electron tunneling from the metal work-function level to the LUMO level of tetracene. This first prototype OLET device was intensively investigated in order to optimize the device characteristics and to explore the possibility of achieving ambipolar transport in tetracene thin films. However, electron currents were never measured in tetracene-based OLETs, which are therefore unipolar OLET devices. In unipolar OLET devices, charges of only one sign are transported along the channel, and the EL emission is localized in a narrow region in the proximity of the electrode that injects the minority carriers. However, most of the scientific and technological characteristics that make light-emitting transistors attractive are only present in ambipolar OLETs for example, the effective formation of a pn junction within the channel with consequent voltage control of exciton formation and radiative recombination.

A thorough description of the effect of each functional component comprising OLET devices is given in Chapter 4. Here, it is worth noting that, in unipolar devices, although an enhancement in EL emission was reached by reducing the device channel length, the emission efficiency is typically low because the exciton radiation and the light extraction are affected by the proximity of the metal contact.

It is therefore evident that the active material properties are essential to achieve a highly performing OLET device. On the one hand, it is rather difficult to find candidates from well-established OLED materials, since most OLED materials do not provide efficient FET charge transport, mainly due to their amorphous morphologies and molecular packing. On the other hand, implementing high-performance OFET active materials typically results in poorly luminescent devices since the tight molecular binding that enables high mobility values is responsible for the activation of exciton nonradiative decay paths.

Most of the single-layer organic field-effect light-emitting transistors that have been fabricated to date show only unipolar conduction (either holes or electrons) even though, in principle, pure organic semiconductor should support both electron and hole transport equally [6]. Meijer *et al.* [33] argued that both hole conduction and electron conduction are generic properties of organic semiconductors that depend on the matching of the energy levels between the HOMO and LUMO of the molecule and the injecting metal work

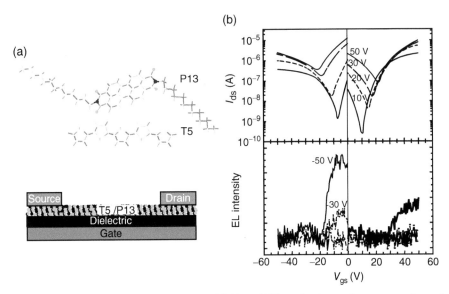

FIGURE 3.9 (a) Molecular structure of P13 and T5, and schematic cross section of the FET device based on the coevaporated thin film of P13 and T5. (b) Transfer characteristics, showing drain current (I_{ds}) and corresponding EL intensity, of a transistor with a T5/P13 ratio of 1:1. [*Source*: Loi *et al.* [38]. Adapted with permission from Wiley.]

functions. However, since the range of metal work functions is limited, in practice, most organic semiconductors will show one conduction polarity.

Besides injection, the use of trap-free dielectrics was mandatory to achieve similar transport properties for both charge carrier types. Electron trapping can be ascribed to the hydroxyl groups (Si–OH) normally present on the silicon dioxide surface [34]. Chua *et al.* [10] demonstrated that the use of an appropriate hydroxyl-free gate dielectric can yield n-channel field-effect conduction in conjugated polymers, which were known to be p-type. It is worth mentioning that the highest reported mobility values for ambipolar transport in the case of small-molecule evaporated films are $0.01/0.1 \, cm^2/V \, s$ (p-type and n-type values, respectively, in carbonyl-functionalized quaterthiophene [35]) and $0.03 \, cm^2/V \, s$ for electron transport in the case of ambipolar solution-processed conjugated polymer (polyselenophene derivative [36]).

Given the limited number of electroluminescent materials with good ambipolar mobility values, different device architectures are to be implemented to achieve high ambipolar transport in OLETs.

Rost *et al.* [37] reported the first ambipolar light-emitting transistor based on coevaporated PTCDI-$C_{13}H_{27}$ (P13) and α-quinquethiophene (α-5T) (Figure 3.9a). In this bulk heterojunction approach, the active layer is formed by the mixing of two unipolar materials with complementary n-transport (P13) and p-transport (α-5T) properties [38]. T5 and P13 are known as good hole- and electron-transporting materials, respectively [31,39], with mobility values of the order of 10^{-2} cm^2/Vs. In this device configuration, exciton formation and charge transport are competitive processes due to the dispersed interface between the p-type and n-type transport materials. Clearly, the wider the interface surface is, the higher the probability that electrons and holes recombine forming excitons. Nevertheless, given that connected percolative paths are needed for the charges to migrate by hopping, the phase interface can represent a physical obstacle for efficient charge transport.

Moreover, the two materials must comply with the relative positions of their HOMO and the LUMO levels in order to allow exciton formation and recombination in the material with the smaller energy gap. In the P13/α-5T bulk heterojunction with a 1:1 coevaporation ratio, light emission is observed for several voltage conditions, and the light intensity is proportional to the drain current (Figure 3.9b).

The controlled coevaporation of α-5T and P13 allows fine-tuning of both the electron and the hole field-effect mobility over two orders of magnitude (10^{-4} to 10^{-2} cm^2/V s). When there is an excess of α-5T, ambipolar transport takes place but no light is detected, which is attributed to quenching of PTCDI-C$_{13}$H$_{27}$ excitons upon interaction with α-5T. For an excess of PTCDI-C$_{13}$H$_{27}$, on the other hand, only n-channel behavior is observed. Nevertheless, light is emitted from the transistor. Both ambipolar transport and light emission are achieved for a balanced ratio of the two components [38]. However, the absolute mobility values are invariably low, of the order of 10^{-3} cm^2/V s.

The observation of a narrow zone of light emission that moves across the channel by changing the gate voltage visualizes and unambiguously proves the existence of two separate channels of holes and electrons in series within the transistor channel in the ambipolar regime. The analytical models proposed to describe single-layer ambipolar OLET and their correlation with experimental data are discussed in Chapter 5. The first material that allowed the observation of the scanning of the light recombination zone in the channel of the transistor in single-layer small-molecule OLETs was a thienopyrrolyl-dione-substituted oligothiophene (T4DIM) (Figure 3.10).

Electrical characterization of T4DIM, which showed ambipolar behavior, was carried out in a bottom gate–top contact geometry OTFT with gold electrodes, under a nitrogen-controlled atmosphere, by using as active layer a 30-nm-thick film of T4DIM grown by vacuum sublimation on top of a glass/ITO/PMMA substrate [40].

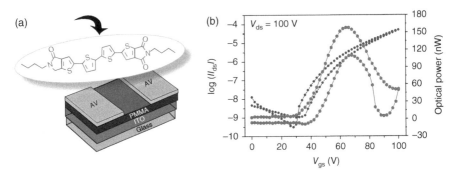

FIGURE 3.10 (a) Molecular structure of the NT4N compound together with the device schematic; (b) n-type transfer characteristic and emitted power curve of the NT4N-based device [*Source*: Melucci *et al.* [40]. Adapted with permission from Royal Society of Chemistry.]

The observation of a spatially resolved recombination zone across the transistor channel on changing the gate bias was first reported in polymer-based ambipolar light-emitting transistors [24,25]. In order to balance the electron and hole injection in the polymer-based ambipolar OLETs, the "two-color" device geometry was implemented, that is, high-work-function metal for the electron-injection electrode and low-work-function metal for the hole-injection electrode. Moreover, a trap-free dielectric was employed in order to enhance the electron transport [10].

Zaumseil *et al.* [41] fabricated BC–TG ambipolar OLETs with poly(9,9-di-*n* octylfluorene-*alt*benzothiadiazole) (F8BT) and poly((9,9-dioctylfluorene)-2, 7-diyl-*alt*-[4,7-bis(3-hexylthien-5-yl)-2,1,3-benzothiadiazole]-2′,2″-diyl) (F8TBT). When operated in the ambipolar regime, F8BT OLETs showed stable, bright green light emission from within the channel. In Figure 3.11a, the position of the emission zone within a 100 μm channel during a scan with fixed drain voltage and variable gate voltage is reported.

Considering the electrical transfer and the corresponding light output characteristics of the device, Figure 3.11b, two peaks in light output versus gate voltage appear when either hole or electron current dominates. The maximum efficiency (0.75%) of this device occurs, as expected, in correspondence with the drain-current minimum. This value is somewhat better than that of the corresponding single-layer F8BT light-emitting diodes (~0.5%), despite partial reflection and absorption of the emitted light by the gold gate.

It was common knowledge that the main drawback affecting the efficiency of ambipolar polymer-based OLETs was related to the low values of charge-carrier mobility (10^{-3} to 10^{-4} cm²/V s). Indeed, a study by Zaumseil *et al.* [23] pointed to a theoretical limit for EQE in a single-layer polymer-based

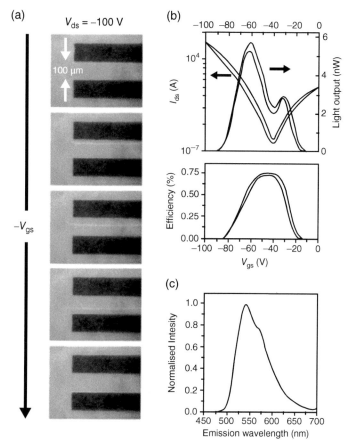

FIGURE 3.11 (a) Optical images of the recombination and emission zones of an F8BT transistor with PMMA as the dielectric during a transfer scan at $V_{ds} = -100V$ and different V_{gs}. (b) Source–drain current is collected by varying V_{gs} from 0 to $-100V$ while keeping V_{ds} at $-100V$ (transfer characteristic) in an F8BT transistor with T-shaped electrodes and PMMA as the gate dielectric. The corresponding light output and external quantum efficiency are also collected. The emission efficiency reaches a maximum plateau (0.75%) when the recombination zone is expected to be located within the channel. (c) Electroluminescence spectrum of a light-emitting F8BT transistor. [*Source*: Zaumseil *et al.* [41]. Adapted with permission from Wiley.]

ambipolar light-emitting transistor. It was demonstrated that in F8BT OLETs, in the top gate configuration, EQE increases with the current density before saturating at around 0.8% due to the trap-assisted mechanism of nonradiative recombination. This value is consistent with the complete recombination of all charges when using a singlet emitter semiconductor. Indeed, it was recently reported that this value can be overcome by one order of magnitude by optimizing the light outcoupling and charge injection [28].

It is evident that EQE in OLETs has to be drastically improved in order to develop technologically relevant applications and to validate OLETs as competitive micrometer-scale light sources with respect to other organic optoelectronic devices, such as OLEDs. In particular, it is necessary to develop an OLET device architecture in which the deleterious photon losses and exciton quenching mechanisms inherent with the OLED structure are suppressed.

In OLED structures, where organic thin-film stacks are sandwiched between an anode and a cathode, the close proximity to the contacts renders metal-induced luminescence quenching important. As we have already shown, truly ambipolar OLET devices allow for a recombination region in the transistor channel far away from the contacts. Consequently, metal-induced quenching becomes irrelevant.

Another essential limitation of the EQE is related to the exciton–polaron quenching of the luminescence [42,43]. This occurs in OLEDs because a substantial number of injected carriers and generated excitons are confined within a very thin layer of the organic semiconductor. In single-layer OLETs, exciton–polaron quenching is expected to remain important because the regions of carrier accumulation and exciton generation largely overlap.

In order to overcome the limitations in optoelectronic performances of single-layer OLETs, organic multilayer structures have been employed as charge-transporting and light-emitting components in OLET devices.

3.3.2 Multilayer OLETs

A possible variation of the coevaporated single-layer structure is to deposit sequentially the two unipolar materials in a vertical stack [44]. Dinelli *et al.* [45] reported that bilayers of α,ω-dihexylquarterthiophene (DH4T) and PTCDI-$C_{13}H_{27}$ show good ambipolar transistor behavior with mobility values as high as 3×10^{-2} cm^2/Vs and light emission (Figure 3.12a).

It was demonstrated that the highest mobility and the more balanced transport are achieved with DH4T sublimed in direct contact with the dielectric. In the OLET bilayer structure, hole and electron transport takes place at the interface between the first organic and the dielectric and at the interface between the two organic layers (Figure 3.12b and c). Morphological analysis of the outermost and buried layers, performed by laser-scanning confocal microscopy, shows that "growth compatibility" between n-type and p-type materials is essential in forming a continuous interface and in controlling the resulting optoelectronic response of the OLET.

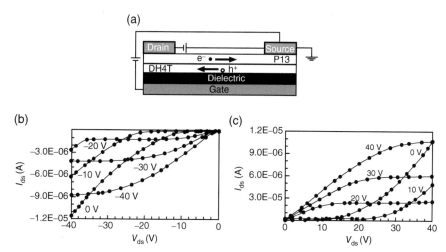

FIGURE 3.12 (a) Schematic of an OLET device based on a DH4T–P13 bilayer. (b) Output curves (I_{ds} vs V_{ds}) obtained from a DH4T–P13 device operating in a p-type configuration or (c) in an n-type configuration. V_{gs} values are reported beside each curve. [*Source*: Dinelli *et al*. [45]. Adapted with permission from Wiley.]

However, the device architecture does not offer any control of the exciton quenching and photon losses as light emission took place only in the unipolar regime. This clearly indicates that light emission was generated underneath the minority-charge-injecting electrodes and not within the channel.

Another strategy for improving OLET light-emission performance is to implement a bilayer film comprising a charge-transport layer and a light-emitting layer in the device active region. It was demonstrated [46] that in devices obtained by spin-casting a polyphenylene vinylene derivative light-emitting polymer (Super Yellow) onto poly-(2,5-bis(3-tetradecylthiophen-2-yl)thieno[3,2-b]thiophene) (PBTTT-C14) hole-transport polymer, the brightness was gate-controlled and exceeded 2500 cd/m^2. The OLETs were fabricated in the bottom gate architecture with asymmetric top contacts Ca/Ag as source/drain electrodes (Figure 3.13).

Recently, the single-layer- and bilayer-based organic structures have been replaced by a trilayer stack consisting of an emission layer sandwiched between an electron- and a hole-transporting layer on a polymeric poly(methyl methacrylate) insulator (Figure 3.14a) [47]. Trilayer OLET devices reach EQE values as large as 5%, outperforming the efficiency of OLEDs with the same emission layer and optimized transport layers (2.2%).

The resulting high EQE is attributed to the active-layer structure that enables a good degree of suppression of essential loss channels for electroluminescence.

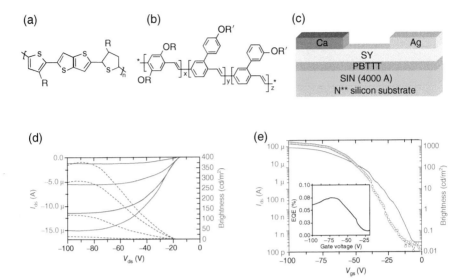

FIGURE 3.13 (a) Molecular structures of PBTTT-C_{14} (R=$C_{14}H_{29}$) and (b) SY. (c) Schematic diagram of the device architecture. (d) Electrical (blue) and optical (red) output characteristic of the OLET. The source–drain voltage was scanned from 0 to −100 V at various gate voltages. (e) Electrical (blue) and optical (red) transfer characteristics of the OLET. The gate voltage was scanned from 0 to −100 V while keeping the source–drain voltage at fixed value of −100 V. Inset shows EQE versus gate voltage. [*Source*: Namdas *et al.* [46]. Adapted with permission of American Institute of Physics.]

It is worth mentioning that other groups have tried to fabricate OLETs implementing similar device architectures, without obtaining comparable optoelectronic performances [48,49].

The main focus in the fabrication of this structure is to enable charges to be injected and diffuse into the central emission layer. So, the LUMO level of the n-type transport layer should be equal to or higher than the LUMO level of the emitting species in the emission layer, while the HOMO level of the p-type transport layer should be lower or aligned with the emitting species HOMO level (Figure 3.14b). Moreover, the control over the interface morphology is mandatory, in order to allow the formation of the most continuous layer stack and high field-effect mobility in the transport layers.

The materials that best fitted these strict requirements are DH4T for p-type transport, α,ω-diperfluorohexylquaterthiophene (DFH4T) for n-type transport, and a blend of the host tris-(8-hydroxyquinoline)aluminum (Alq_3) and guest 4-(dicyanomethylene)-2-methyl-6-(*p*-dimethylaminostyryl)-4H-pyran (DCM) for the emission layer (Figure 3.14a)

(a)

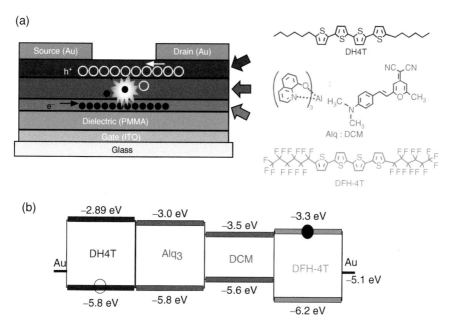

(b)

FIGURE 3.14 (a) Schematic representation of the trilayer OLET device with the chemical structure of each material making up the device active region. The field-effect charge transport and the light-generation processes are also sketched. (b) Energy-level diagram of the trilayer heterostructure. The energy values of the HOMO and LUMO levels of each molecular material are indicated together with the Fermi level of the gold contacts. [*Source*: Capelli *et al.* [47]. Adapted with permission from Nature Publishing Group.]

When the device is operated in unipolar regime ($|V_{ds}|=|V_{gs}|$) in n-polarization (Figure 3.15a), a diode-like mechanism gives rise to light emission in the proximity of the drain contact, as it is observed from the linear correlation between the EL and the current intensity. In Figure 3.15b, the electrical-transfer and light-output characteristics when the applied bias conditions allow simultaneous charge injection from both the source and drain electrodes ($|V_{ds}|<|V_{gs}|$) are reported.

Even though the unbalanced transport in the heterostructure hinders the V-shaped characteristic typical of ambipolar transistor in transfer mode, a new mechanism in light generation is clearly visible by comparing the light output in unipolar (Figure 3.15a) and ambipolar (Figure 3.15b) regimes. Indeed, the shaded area in Figure 3.15b highlights the electroluminescence produced by the ambipolar current. Moreover, in the ambipolar operation mode ($V_{ds}=100\,V$ and $V_{gs}=30\,V$), the light-emitting region is located far from the drain electrodes, at a distance of $8\,\mu m$, thus preventing photon losses due to exciton–metal quenching (see Section 5.3).

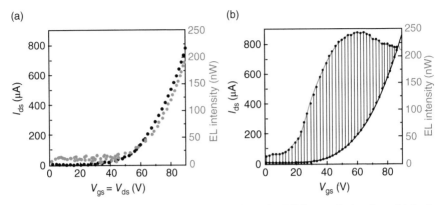

FIGURE 3.15 (a) Locus electrical curves of the OLET in n-polarization. (b) In the transfer characteristic curves, the source–drain current (I_{ds}) is measured keeping the drain–source potential constant at 90 V, while sweeping the gate–source potential from 0 to 90 V. [*Source*: Capelli *et al.* [47]. Adapted with permission from Nature Publishing Group.]

From the engineering point of view, it is interesting to note that the structure can be inverted to have either the n-type or the p-type semiconductor in contact with the dielectric. This offers important degrees of freedom for the design of optimized device architectures that take into account the physical properties and the processing requirements of the materials to be employed in the active region of the device. In three-layer OLETs, the different locations of the light-emitting area and the charge-carrier accumulation zones separate the regions of large carrier densities from the regions of large exciton densities, reducing exciton–polaron quenching. The avoidance of exciton quenching and photon losses phenomena was experimentally demonstrated by the fact that an optimized OLED with the same emitter layer presented an EQE more than halved with respect to the optimized trilayer OLET (for details, see Section 6.1).

Although many fundamental aspects of the correlation between the structure and mechanisms of charge transport at solution-processed organic/organic heterointerfaces remain to be explored in order to obtain OLETs with overall transistor performances (bias voltage values, hysteresis, reliability) competitive with those of vacuum-sublimed field-effect devices [50], few significant reports have appeared in the literature. An OLET comprising a trilayer film consisting of a hole-transporting layer, an emissive layer, and a conjugated polyelectrolyte (CPE) layer acting as an electron injection layer has been reported [51] (Figure 3.16).

The device has symmetric high-work-function source and drain metal electrodes and demonstrates that a solution-processed fabrication methodology is possible and has the potential to be optimized to achieve multicolor OLETs.

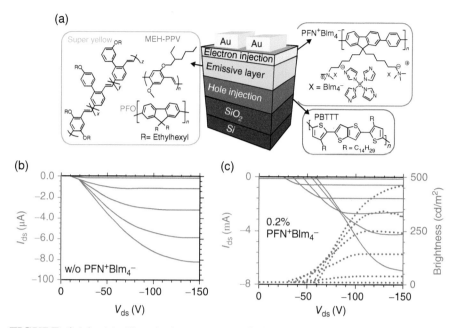

FIGURE 3.16 (a) Chemical structures of the materials implemented in the solution-processed trilayer OLET with the Si/SiO$_2$/OTS/PBTTT/MEH-PPV/PFN$^+$BIm$_4^-$/Au device configuration. (b) Electrical output (blue line) characteristics of the control device OLET configuration (without electron-injecting and -transporting layer). (c) Electrical output (blue line) and optical output (red circle) characteristics of the best-performing solution-processed trilayer OLET implementing electron-injecting an transporting layer deposited from 0.2% solution on top of the MEH-PPV layer. [*Source*: Seo *et al.* [51]. Adapted with permission from Wiley.]

It is important to note that the introduction of the trilayer heterostructure approach in OLET technology offers a much wider flexibility in the fabrication and restricts the limitation of the light-emitting field-effect technology to only material-related issues.

3.4 SINGLE-CRYSTAL OLET

With the advent of devices based on single crystals, the performance of organic FETs has experienced a significant leap, with mobility now in excess of 10 cm^2/Vs in the case of rubrene single crystal [7].

Using single crystals makes it easier to analyze the effects of defects and impurities on the device characteristics and to get closer to the limits of the intrinsic electrical properties [6]. Thin films are not appropriate for this kind of analysis because their morphological and electrical properties are strongly related to the characteristics of the substrate on which they are deposited. As we will show in the next chapter, the insulating layer influences both the growth process and the electrostatic potential in the first layers of the semiconductor, and these two contributions are not easily separable.

Owing to the strong π–π interaction among molecules, the single-crystal growth of organic semiconductor materials for use as active components in field effect devices needs the employment of specially adapted techniques. Indeed, the single-crystal growth of organic semiconductors has a key role in the development of high-performance devices. A straightforward way to make a device is to use a monolithic single crystal. When the monolithic crystal is obtained as a free-standing thin film, it can be laminated on a device substrate. Alternatively, the monolithic crystal can be formed in situ on top of the device substrate. The crystal growth methods include the vapor-phase growth and liquid-phase growth. In the vapor-phase case, the growth method can further be sorted into those in an open system or in a closed one.

The quality of the deposited single crystal (thus, of the optoelectronic characteristics) can be further improved by treating the substrate with various species of self-assembled monolayers, for example, 1,1,1,3,3,3-hexamethyl-disilazane (HMDS), phenyltrichlorosilane (PTS), and octadecyltrichlorosi-lane (OTS). Once the substrate has been functionalized, the crystal can be laminated on the substrate or directly grown on top of the substrate. Without the SAMs, the grown crystal may be subjected to cracks and irregularities originated by the high temperatures needed for the crystal growth. Indeed, cracks are likely to be generated during the crystal growth process because of the difference in thermal expansion coefficients between the organic semi-conductor crystal and the supporting substrate (e.g., Si and SiO_2) [52].

In order to fabricate working transistor devices, the substrate is typically a conductive one and is topped with a thin insulating layer. A typical example is a silicon wafer with a silicon oxide on top, in which case, the wafer main body and silicon oxide constitute the gate electrode and gate insulator, respectively. The electrodes deposited on the single crystal form the source and drain contacts (top contact configuration). Alternatively, both the contacts may be grown on the dielectric layer and located underneath the organic single crystal in a bottom contact configuration. In an additional embodiment, the contacts may be deposited on one side of the single crystal and the dielectric layer and gate contact may be fabricated directly on the other side of the crystal.

Differently from OLETs based on polycrystalline thin films, OLETs presenting single crystal as active layer often require small-work-function metal such as Ca and Mg as electrode for injecting electrons. Indeed, at the early

(a)

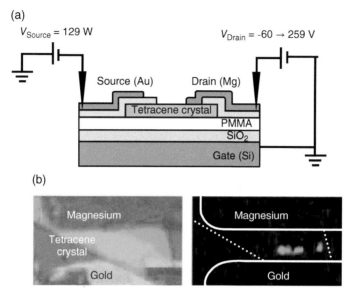

(b)

FIGURE 3.17 (a) Schematic representation of tetracene-based single-crystal OLETs together with applied bias conditions. (b) Images of a tetracene single-crystal OLET in the light (left) and in the dark (right, $V_s = 129\,$V and $V_{ds} = -259\,$V). The white solid and dotted lines are for visual guide only. [*Source*: Takahashi *et al.* [53]. Adapted with permission from Wiley.]

stage of development of OLETs based on single crystals, it was necessary to implement asymmetric electrodes for obtaining light emission from devices based on unsubstituted conjugated organic small molecules [53].

As in the case of OLETs based on polycrystalline thin films, when applying ambipolar bias conditions, the p- and n-type regions are created at the source and drain side, respectively, of the channel. This process leads to the formation of a forward-biased p-n junction within the device, and consequently, the generation of excitons is expected. Although the emission zone position is controlled within the channel by acting on the gate bias as expected, the visible light emission is usually observed at extremely high operating voltages due to the high threshold voltages for injection and transport of holes and electrons.

It is interesting to note that the emission area in single-crystal OLETs is more inhomogeneous than that typically observed in OLETs based on polycrystalline thin films. Possible reasons for this inhomogeneity include issues with the single-crystal lamination on the substrate and less uniform charge-carrier distribution (Figure 3.17). Indeed, the emission zone features are clearly correlated to the charge recombination process within the channel and are in turn correlated to the charge-transport physics in single-crystal OLETs.

Generally, the structure of organic single crystals is obtained from the stacking of well-ordered molecules along a certain direction. So, if properly oriented with respect to the dielectric surface, organic single crystals in a field-effect device structure are expected to show high current density and ambipolar behavior. However, the compact molecular crystal packing can in turn give rise to increasing nonradiative decay paths for excitons, resulting in low luminescence efficiency. Therefore, great effort has been devoted to optimize organic single crystals where high charge mobility and high luminescence efficiency coexist.

In this respect, a recently developed class of materials of thiophene/phenylene cooligomers (TPCOs) [54] fills a special niche in the OLET architecture. These materials exhibit an exceedingly high luminescent quantum efficiency in the form of a crystal. As an example, Kanazawa et al. [55] estimated the fluorescent quantum efficiency of crystals of α,ω-bis(biphenylyl)terthiophene (B3PT, one of TPCOs) to be 80% at 300 K. This value is particularly impressive if compared with the value of 1% observed for the emission efficiency of single crystals of rubrene and tetracene [56].

Several reports demonstrated that BP3T in the form of amorphous thin films can be used in ambipolar OLETs devices [57] while single-crystal devices are unipolar [58], the latter having hole mobilities of the order of $10^{-1}\,cm^2/V\,s$. Recently, electron transport in BP3T-based devices was promoted by inserting a buffer layer of PMMA between the crystal and the SiO_2 dielectric layers [56] (Figure 3.18a). Indeed, the hole and electron mobility values achievable after the surface treatment are 1.64 and $0.17\,cm^2/Vs$, respectively (Figure 3.18b).

In general, the thiophene/phenylene co-oligomers molecules substituted with both electron-donating and electron-withdrawing groups show ambipolar characteristics, as is the case of 2-{4-[5-(4-methoxyphenyl)-2-thienyl] phenyl}-5-[4-(trifluoromethyl)phenyl]thiophene (AC5-1CF3-12OMe) [59]. These features render the TPCO materials particularly attractive for light-emitting field-effect devices, due to their combination of high brightness and efficient field-effect transport.

In the case of linear conjugated molecules presenting large dipole moment along the molecular long axis, such as TPCO, it is possible to obtain the most intense light emission along the direction perpendicular to the molecular long axis. Usually, the molecules are arranged nearly perpendicular to the ab crystal plane [60] in H-aggregate-like stacking so that the molecular emission tends to be directed parallel to the plane. As a consequence, electroluminescence emission from the edges of the crystal dominates with respect to the vertical emission. Moreover, the OLET architecture is advantageous for the confinement of light, which is emitted within small diverging angles. This is

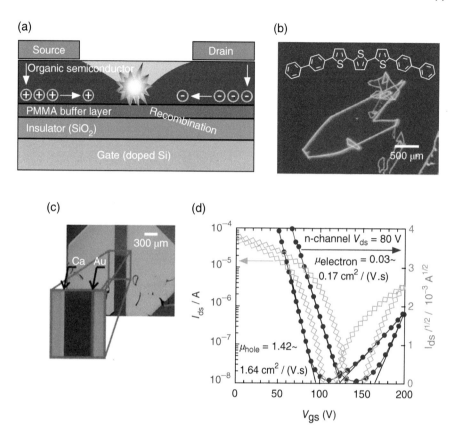

FIGURE 3.18 (a) Schematic diagram of the single-crystal ambipolar light-emitting transistor architecture and operation. (b) An optical micrograph of the fabricated LET device with Au and Ca asymmetric electrodes (length 240 µm, width 1320 µm). (c) Plots of transfer characteristics (open squares) of the device during n-channel operation at $V_{ds} = 80$V, along with the corresponding $I_{ds}^{1/2} - V_{gs}$ plots (solid circles), from which the carrier mobilities were extracted. [*Source*: Bisri *et al.* [56]. Adapted with permission from Wiley.]

originated by the small tilting angle of the molecules relative to the *ab* crystal plane normal axis together with the nearly upright configuration of the transition dipole moments. With reference to Figure 3.19, in the case of BP3T single crystal, the slab-stack waveguide structure of N_2/single crystal/SiO_2 causes any diverging light having a small grazing angle to be totally internally reflected between the two slab interfaces.

This structure prevents the emission leaking out from the crystal surface and makes it self-waveguided, so that it is eventually edge-emitted. Henceforth, it gives the possibility of collecting all of the electroluminescence emission at very narrow single (or double) exits, which are the crystal (channel) edges. The light is expected to be emitted in all directions in the 2D

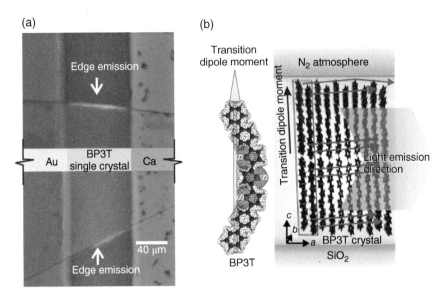

FIGURE 3.19 (a) Edge emission of BP3T single-crystal ambipolar LET ($L = 80 \mu m$, $W = 860 \mu m$) during ambipolar operation, as observed from above for identical device edges under ambient light conditions. (b) A schematic illustration of the self-waveguided edge-emission mechanisms, showing the BP3T molecules stacked nearly perpendicular to the crystal plane with their transition dipole moments and molecular long axis in parallel. Thus, it causes a tendency for the molecular light emission to be parallel to the crystal plane. The OLET structure itself forms a BP3T crystal-cored waveguide. [*Source*: Bisri *et al.* [56]. Adapted with permission from Wiley.]

crystal plane from its origin in the recombination zone. Therefore, the edge emission itself is supposed to be observed from the whole crystal edge, with intensity maximum in correspondence of the recombination zone.

Interestingly, the edge emission prompts spectrally narrowed polarized emissions as the drain current increases in the ambipolar regime. This particular behavior is due to the regular molecular alignment, which produces a large optical gain in the direction parallel to the wide crystal plane. As a consequence, the amplification of self-waveguided propagating light is readily attained [61]. The possible mechanisms at the basis of this phenomenon may be ascribed to (i) cut-off mode [62], (ii) leaky mode, [63], or (iii) amplified spontaneous emission (ASE) or stimulated emission. Among these three possible mechanisms, none is conclusive yet for explaining the optoelectronic and photonic properties of single-crystal OLETs based on TPCO molecules. The possibility of having spectral narrowing via leaky mode might be ruled out: indeed, the leaky mode propagating out of the waveguide core into the underneath SiO_2 dielectric layer would be absorbed by the doped Si substrate because of its metallic behavior.

The hypothesis of an initial emission-gain narrowing is corroborated by the current-density-dependent spectral change, which is experimentally observed. In this respect, we note that the small optical loss correlated to crystal slab-stack waveguide may cause the spectrally narrowed emissions, irrespective of what the origin of the emissions is (i.e., ASE or stimulated emission) [64]. The self-waveguiding might prompt such emissions. Nonetheless, the estimation of the current density needed to achieve the threshold measured for optically pumped ASE is two orders of magnitude higher with respect to the measured OLET current density in ambipolar conditions.

Finally, line narrowing due to the presence of a cut-off mode is a plausible explanation once perfect refraction takes place at the interface between the organic and the substrate. The wavefront that propagates along the interface would be possibly amplified, since it would partially interact with the organic molecules along the interface. This amplification has also been identified to be current-dependent in a supralinear manner [62]. With this mechanism, the wavelength that is mode-selected as cut-off mode is determined by the crystal thickness, the overall waveguide structure, and the crystal refractive index.

Evidently, the discussion on spectral narrowing in single-crystal OLETs is still open to debate, and further investigation has to be conducted in order to assess the real mechanism related to the experimental findings. As we will discuss in detail in Chapter 6, those two possible mechanisms are promising paths for realizing electrically driven laser oscillation from organic materials in single-crystal based OLETs. Interestingly, the optical confinement and the self-waveguiding coupled with Davydov splitting, which is another typical feature of organic crystals, may be suitably exploited in order to obtain red/green/blue emission from an individual single-crystal OLET [65].

Nonetheless, the other necessary condition for achieving lasing in field-effect device architecture is the high current density. Indeed, it was observed the absence of roll-off behavior in EQE up to the current density of several hundred (or thousand) amperes per square centimeter range, which is two orders of magnitude larger than the maximum current density in conventional OLED devices [66]. The efficiency of single-crystal OLETs can be increased by one order of magnitude by dye-doping the organic crystal. The doping of tetracene molecules into p-distyrylbenzene (P3V2) [67] causes an increase of EQE at values of 0.64% (Figure 3.20a), together with an EL color-tuning from P3V2 (host) to tetracene (guest) emission (Figure 3.20b). Interestingly, a slight increase of quantum efficiency with the flowing drain current was found, from a linear regime at low current intensity up to saturation for high current intensity. The reasons for this unexpected behavior are still under debate.

Recently, an investigation on how to increase the current density in single-crystal OLETs by decreasing the electron-injection barrier and adapting the current-confinement structure [68] was carried out. Sawabe et $al.$ [69]

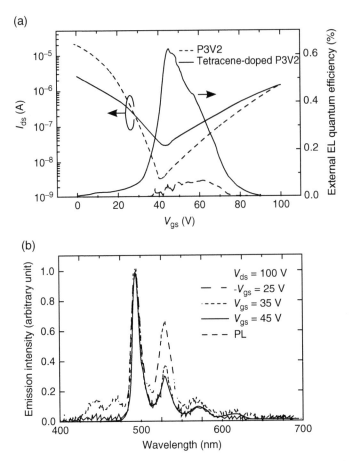

FIGURE 3.20 (a) Transfer characteristics and dependence of external quantum efficiency on V_{gs} in OLETs containing P3V2 and tetracene-doped P3V2 crystals with asymmetric gold–calcium electrodes. (b) PL and EL spectra of an OLET containing a tetracene-doped P3V2 crystal at a constant V_{ds} of 100 V and various V_{gs}. [*Source*: Nakanotani *et al.* [67]. Adapted with permission from Wiley.]

demonstrated that maximum current up to $33 \, kA/cm^2$ (assuming a 3 nm thick accumulation layer) was achieved in single-crystal OLET based on BP3T. Interestingly, the luminescence in the device was observed to increase linearly up to a current density of $1 \, kA/cm^2$.

In order to achieve well-balanced ambipolarity and low electron threshold voltage, the device architecture reported in Figure 3.21a was implemented [68]. The architecture is obtained by incorporating an electron-injection buffer layer (CsF) into the single-crystal OLET based on BP3T to reduce the injection barrier. In addition, a current-confinement structure was embedded

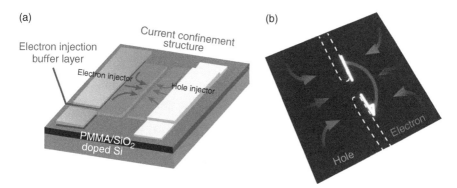

FIGURE 3.21 (a) Design principle for achieving high current density in organic single-crystal light-emitting transistor. An electron-injection buffer layer to decrease the injection barrier is implemented. Moreover, the transistor channel was laser-etched to confine the current pathway. The blue and red arrows indicate the electron and hole currents, respectively. (b) Snapshots of the light emission observed from the laser–organic single-crystal light-emitting transistor. The arrows and dashed lines present a guide for the eye. [*Source*: Sawabe *et al*. [68]. Adapted with permission from Wiley.]

into the device to further enhance the current. The structure was fabricated by using a laser etching technique to narrow the transistor channel so that the current path was limited to the portion of the channel not removed by the etching (width of 10 μm and length of 4 μm). The laser-etched device showed highly balanced ambipolar behavior due to CsF buffer layer. The recombination zone in the channel of the OLET structure with current confinement is not straight but ellipsoidal, given the electric-field distribution in the confinement region (Figure 3.21b). The increased light emission from the laser-etched crystal edges can be explained by the perfect alignment of the transition dipole moments and the self-waveguiding properties of BP3T single crystals, as we have discussed before. As the recombination zone approached the confined region, the length of the ellipsoidal emissions decreased drastically and the emission became brighter, which suggests an increment in the current density. Using the accumulation layer thickness and recombination zone length, a maximum current density of approximately 33 kA/cm² was estimated, which is one order of magnitude higher than the values previously achieved in both single-crystal and polymeric OLETs [46,69].

Finally, it is important to underline that to promote ambipolar charge transport in single-crystal OLETs, Yamao *et al*. [70] have adopted a novel device-operation method, which is characterized by an AC voltage being applied to the gate electrode. This method enables the use of a stable metal such as Au as an electrode for injecting both electrons and holes. They reported an EQE of $3.7 \times 10^{-4}\%$ for a device using a crystal of a thiophene/phenylene co-oligomer.

By using this gate-bias pattern, Kajiwara *et al.* [71] demonstrated bright emissions from devices comprising bilayer organic crystals of a p-type and an n-type. This structure enables the efficient injection and transport of both the electrons and holes, leading to their recombination with a maximum EQE of 0.045%. The device structure combined with the AC-gate-bias operation might offer a promising strategy to achieve bright and efficient light emissions in single-crystal OLETs.

3.5 CONCLUSIONS

In this chapter, the working principles of OLETs were introduced and the key features of the widely investigated light-emitting field-effect device structures were reviewed. The various architectures in which the active region of OLETs has been engineered were described to highlight properties and limitations of each of them. The specificity of single-crystal OLETs was illustrated with particular reference to their fundamental properties that might become especially relevant in the search for injection lasing from organic semiconducting materials. A possible innovative strategy to achieve organic injection lasing, which builds on these fundamental properties, will be outlined in Section 6.4. The key functional building blocks of the OLET device structure are described and discussed in Chapter 4.

REFERENCES

[1] J. A. Rogers, Z. Bao, A. Dodabalapur, B. Crone, V. R. Raju, H. E. Katz, V. Kuck, K. J. Ammundson, and P. Drzaic, *Proc. Natl. Acad. Eng.* **2001**, *98*, 4817.

[2] C. Videlot, J. Ackermann, P. Blanchard, J.-M. Raimundo, P. Frère, M. Allain, R. de Bettignies, E. Levillain, and J. Roncali, *Adv. Mater.* **2003**, *15*, 306.

[3] C. Videlot, A. El Kassmi, and D. Fichou, *Sol. Energy Mater. Sol. Cells* **2000**, *63*, 69.

[4] H. Sirringhaus, P. J. Brown, R. H. Friend, M. M. Nielsen, K. Bechgaard, B. M. W. Langeveld-Voss, A. J. H. Spiering, R. A. J. Janssen, E. W. Meijer, P. Hervig, and D. M. de Leeuw, *Nature* **1999**, *401*, 685.

[5] G. Li, V. Schrotriya, J. Huang, Y. Yao, T. Moriarty, K. Emery, and Y. Yang, *Nat. Mater.* **2005**, *4*, 864.

[6] N. Karl, *Synth. Metal.* **2003**, *133*, 649.

[7] D. Braga and G. Horowitz, *Adv. Mater.* **2009**, *21*, 1473.

[8] M. T. Bernius, M. Inbasekaran, J. O'Brien, and W. S. Wu, *Adv. Mater.* **2000**, *12*, 1737.

[9] H. Becker, H. Spreitzer, W. Kreuder, E. Kluge, H. Schenk, I. Parker, and Y. Cao, *Adv. Mater.* **2000**, *12*, 42.

[10] L. L. Chua, J. Zaumseil, J. F. Chang, E. C. W. Ou, P. K. H. Ho, H. R. Sirringhaus, and H. Friend, *Nature* **2005**, *434*, 194.

[11] S. Kobayashi, T. Nishikawa, T. Takenobu, S. Mori, T. Shimoda, T. Mitani, H. Shimotani, N. Yoshimoto, S. Ogawa, and Y. Iwasa, *Nat. Mater.* **2004**, *3*, 317.

[12] K. P. Pernstich, S. Haas, D. Oberhoff, C. Goldmann, D. J. Gundlach, B. Batlogg, A. N. Rashid, and G. J. Schitter, *Appl. Phys.* **2004**, *96*, 6431.

[13] D. J. Frank, R. H. Dennard, E. Nowak, P. M. Solomon, Y. Taur, and H. S. P. Wong, *Proc. IEEE* **2001**, *89*, 259.

[14] G. S. Tulevski, C. Nuckolls, A. Afzali, T. O. Graham, and C. R. Kagan, *Appl. Phys. Lett.* **2006**, *89*, 183101.

[15] J. N. Haddock, X. H. Zhang, S. J. Zheng, S. R. ZhangMarder, and B. Kippelen, *Org. Electron.* **2006**, *7*, 45.

[16] A. Ortiz-Conde, *et al.*, *Microelectronics Reliability* **2002**, *42*, 583.

[17] J. Veres, S. Ogier, G. Lloyd, and D. De Leeuw, *Chem. Mat.* **2004**, *16*, 4543.

[18] C. R. Newman, C. D. Frisbie, D. A. Da Silva Filho, J. L. Bredas, P. C. Ewbank, and K. R. Mann, *Chem. Mater.* **2004**, *16*, 4436.

[19] R. Schmechel, M. Ahles, and H. Von Seggern, *J. Appl. Phys.* **2005**, *98*, 084511.

[20] I. Kymissis, C. D. Dimitrakopolous, and S. Purushothaman, *IEEE Trans. Elec. Dev.* **2001**, *48*, 1060.

[21] C. D. Dimitrakopolous, A. R. Brown, and A. Pomp, *J. Appl. Phys.* **1996**, *80*, 2501.

[22] P. V. Pesavento, *et al.*, *J. Appl. Phys.* **2004**, *96*, 7312.

[23] J. Zaumseil, C. R. McNeill, M. Bird, D. L. Smith, P. Ruden, M. Roberts, M. J. McKiernan, R. H. Friend, and H. Sirringhaus, *J. Appl. Phys.* **2008**, *103*, 064517.

[24] J. Zaumseil, R. H. Friend, and H. Sirringhaus, *Nature Mater.* **2006**, *5*, 69.

[25] J. S. Swensen, C. Soci, and A. J. Heeger, *Appl. Phys. Lett.* **2005**, *87*, 253511.

[26] B. B. Y. Hsu, C. Duan, E. B. Namdas, A. Gutacker, J. D. Yuen, F. Huang, Y. Cao, G. C. Bazan, I. D. W. Samuel, and A. J. Heeger, *Adv. Mater.* **2012**, *24*, 1171.

[27] B. B. Y. Hsu, E. B. Namdas, J. Yuen, I. D. W. Samuel, and A. J. Heeger, *Adv. Mater.* **2010**, *22*, 4649.

[28] M. C. Gwinner, D. Kabra, M. Roberts, T. J. K. Brenner, B. H. Wallikewitz, C. R. McNeill, R. H. Friend, and H. Sirringhaus *Adv. Mater.* **2012**, *24*, 2728.

[29] S. Schols, S. Verlaak, C. Rolin, D. Cheyns, J. Genoe, and P. Heremans, *Adv. Funct. Mater.* **2008**, *18*, 136.

[30] A. Hepp, H. Heil, W. Weise, M. Ahles, R. Schmechel, and H. von Seggern, *Phys. Rev. Lett.* **2003**, *91*, 157406.

[31] P. R. L. Malenfant, C. D. Dimitrakopoulos, J. D. Gelorme, A. Curioni, and W. Andreoni, *Appl. Phys. Lett.* **2002**, *80*, 2517.

[32] C. Santato, R. Capelli, M. A. Loi, M. Murgia, F. Cicoira, V. A. L. Roy, P. Stallinga, R. Zamboni, C. Rost, S. F. Karg, and M. Muccini *Synth. Met.* **2004**, *146*, 329.

[33] E. J. Meijer, D. M. de Leeuw, S. Setayesh, E. van Veenendaal, B.-H. Huismna, P. W. M. Blom, J. C. Hummelen, U. Scherf, and T. M. Klapwijk, *Nature Mater.* **2003**, *2*, 678.

[34] F. Todescato, R. Capelli, F. Dinelli, M. Murgia, N. Camaioni, M. Tang, R. Bozio, and M. Muccini, *J. Phys. Chem. B* **2008**, *112*, 10130.

[35] M. H. Yoon, S. A. DiBenedetto, A. Facchetti, and T. Marks, *J. Am. Chem. Soc.* **2005**, *127*, 1348.

[36] Z. Chen, H. Lemke, S. Albert-Seifried, M. Caironi, M. Meedom Nielsen, M. Heeney, W. Zhang, I. McCulloch, and H. Sirringhaus, *Adv. Mater.* **2010**, *22*, 2371.

[37] C. Rost, S. Karg, W. Riess, M. A. Loi, M. Murgia, and M. Muccini, *Appl. Phys. Lett.* **2004**, *85*, 1613.

[38] M. A. Loi, C. Rost-Bietsch, M. Murgia, S. Karg, W. Riess, and M. Muccini, *Adv. Funct. Mater.* **2006**, *16*, 41.

[39] M. Melucci, M. Gazzano, G. Barbarella, M. Cavallini, F. Biscarini, P. Maccagnani, and P. Ostoja, *J. Am. Chem. Soc.* **2003**, *125*, 10266.

[40] M. Melucci, M. Zambianchi, L. Favaretto, M. Gazzano, A. Zanelli, M. Monari, R. Capelli, S. Troisi, S. Toffanin, M. Muccini, *Chem. Commun.* **2011**, *47*, 11840.

[41] J. Zaumseil, C. L. Donley, J.-S. Kim, R. H. Friend, and H. Sirringhaus, *Adv. Mater.* **2006**, *18*, 2708.

[42] H. Yamamoto, T. Oyamada, H. Sasabe, and C. Adachi, *Appl. Phys. Lett.* **2004**, *84*, 1401.

[43] W. W. A. Koopman, S. Toffanin, M. Natali, S. Troisi, R. Capelli, V. Biondo, A. Stefani, M. Muccini, *Nano Lett.* **2014**, *14*, 1695.

[44] C. Rost, S. Karg, W. Riess, M. A. Loi, M. Murgia, and M. Muccini, *Synth. Met.* **2004**, *146*, 237.

[45] F. Dinelli, R. Capelli, M. A. Loi, M. Murgia, M. Muccini, A. Facchetti, and T. J. Marks, *Adv. Mater.* **2006**, *18*, 1416.

[46] E. B. Namdas, P. Ledochowitsch, J. D. Yuen, D. Moses, and A. J. Heeger, *Appl. Phys. Lett.* **2008**, *92*, 183304.

[47] R. Capelli, S. Toffanin, G. Generali, H. Usta, A. Facchetti, and M. Muccini, *Nature Mater.* **2010**, *9*, 496.

[48] V. Maiorano, A. Bramanti, S. Carallo, R. Cingolani, and G. Gigli, *Appl. Phys. Lett.* **2010**, *96*, 133305.

[49] E. B. Namdas, I. D. W. Samuel, D. Shukla, D. M. Meyer, Y. Sun, B. B. Y. Hsu, D. Moses, and A. J. Heeger, *Appl. Phys. Lett.* **2010**, *96*, 043304.

[50] F. Cicoira and C. Santato, *Adv. Func. Mater.* **2007**, *17*, 3421.

[51] J. H. Seo, E. B. Namdas, A. Gutacker, A. J. Heeger, and G. C. Bazan, *Adv. Funct. Mater.* **2011**, *21*, 3667.

[52] J. S. Brooks, *Chem. Soc. Rev.* **2010**, *39*, 2667.

[53] T. Takahashi, T. Takenobu, J. Takeya, and Y. Iwasa, *Adv. Funct. Mater.* **2007**, *17*, 1623–1628.

[54] S. Hotta and T. Katagiri, *J. Heterocycl. Chem.* **2003**, *40*, 845.

[55] S. Kanazawa, M. Ichikawa, T. Koyama, and Y. Taniguchi, *ChemPhysChem*, **2006**, *7*, 1881.

[56] S. Z. Bisri, T. Takenobu, Y. Yomogida, H. Shimotani, T. Yamao, S. Hotta, and Y. Iwasa, *Adv. Funct. Mater.* **2009**, *19*, 1728.

[57] K. Yamane, H. Yanagi, A. Sawamoto, S. Hotta, *Appl. Phys. Lett.* **2007**, *90*, 162.

[58] K. Nakamura, M. Ichikawa, R. Fushiki, T. Kamikawa, M. Inoue, T. Koyama, and Y. Taniguchi, *Jpn J. Appl. Phys.* **2005**, *44*, L1367.

[59] T. Katagiri, Y. Shimizu, K. Terasaki, T. Yamao and S. Hotta, *Org. Electron.* **2011**, *12*, 8.

[60] S. Hotta, M. Goto, R. Azumi, M. Inoue, M. Ichikawa, and Y. Taniguchi, *Chem. Mater.* **2004**, *16*, 237.

[61] S. Hotta, M. Goto and R. Azumi, *Chem. Lett.* **2007**, *36*, 270.

[62] D. Yokoyama, H. Nakanotani, Y. Setoguchi, M. Moriwake, D. Ohnishi, M. Yahiro, and C. Adachi, *Jpn. J. Appl. Phys.* **2007**, *46*, L826.

[63] Y. Tian, Z. Q. Gan, Z. Q. Zhou, D. W. Lynch, J. Shinar, J. H. Kang, and Q. H. Park, *Appl. Phys. Lett.* **2007**, *91*, 143.

[64] K. L. Shaklee and R. F. Leheny, *Appl. Phys. Lett.* **1971**, *18*, 475.

[65] Y. Yomogida, H. Sakai, K. Sawabe, S. Gocho, S. Z. Bisri, H. Nakanotani, C. Adachi, T. Hasobe, Y. Iwasa, and T. Takenobu, *Organic Electronics* **2013**, *14*, 2737.

[66] T. Takenobu, S. Zulkarnaen Bisri, T. Takahashi, M. Yahiro, C. Adachi, and Y. Iwasa, *Phys. Rev. Lett.* **2008**, *100*, 066601.

[67] H. Nakanotani, M. Saito, H. Nakamura, and C. Adachi, *Adv. Funct. Mater.* **2010**, *20*, 1610.

[68] K. Sawabe, M. Imakawa, M. Nakano, T. Yamao, S. Hotta, Y. Iwasa, and T. Takenobu, *Adv. Mater.* **2012**, *24*, 6141.

[69] K. Sawabe, T. Takenobu, S. Z. Bisri, T. Yamao, S. Hotta, and Y. Iwasa, *Appl. Phys. Lett.* **2010**, *97*, 043307.

[70] T. Yamao, Y. Shimizu, K. Terasaki, and S. Hotta, *Adv. Mater.* **2008**, *20*, 4109.

[71] K. Kajiwara, K. Terasaki, T. Yamao, and S. Hotta, *Adv. Funct. Mater.* **2011**, *21*, 2854.

4

KEY BUILDING BLOCKS OF OLETs

In this chapter, the main constituting elements of the organic light-emitting transistor (OLET) device are introduced and their key fundamental features analyzed. The role and the effect of different dielectric materials on device performance are discussed and related to their structural, morphological, and chemical properties. The requested characteristics for the active layer in an OLET are discussed, and practical examples of OLETs based on polymeric and molecular ambipolar emissive materials are reviewed. The influence of the charge-injecting electrodes on the OLET characteristics is highlighted, with particular emphasis on the role played by the interface between metals and organic semiconductors (OSCs).

4.1 DIELECTRIC LAYER

The dielectric layer in a typical thin-film transistor device is sandwiched between the gate electrode and the OSC.

This geometry creates two interfaces, which must be considered when selecting a suitable dielectric material for OLETs. When considering the dielectric–gate interface, one must address the issue of preventing static charge or dynamic charge injection into the dielectric, which can have adverse effects on threshold voltage (V_t) as a function of time. Furthermore, complete coverage of the gate electrode is necessary to prevent leakage currents through pinhole defects.

Organic Light-Emitting Transistors: Towards the Next Generation Display Technology, First Edition. Michele Muccini and Stefano Toffanin.

The dielectric–semiconductor interface is where the conducting channel is formed so that the control of the interface roughness, surface energy, and charge accumulation at this interface is fundamental in mastering the optoelectronic performances in OLETs [1]. Furthermore, in most thin-film transistor structures, the semiconductor is grown or deposited on top of the dielectric. In these instances, the structure of the first few monolayers of the OSCs at the interface with the dielectric plays a major role in allowing the field-effect charge transport and eventually electroluminescence [2,3].

The key features to be considered when choosing a gate dielectric for an organic transistor are control of interfaces, leakage, dielectric constant, processability, stability, and reliability.

Achieving high capacitance is of great importance, considering the low mobilities (typically $<1 \, cm^2/V \, s$) of OSCs [1,4,5]. Keeping leakage currents at a minimum will reduce the power requirements of these devices [6] and prevent early-stage failure. Ease of processing of dielectric materials is an important issue when evaluating the manufacturing in view of commercialization. Finally, the dielectric films must be stable (i.e., not prone to drifts in performance or threshold voltage) and exhibit minimum hysteresis.

In the case of single-layer OLETs, the dielectric layer has to be chosen so as to promote and enhance the ambipolar charge transport within the polymeric or small-molecule OSC and to allow efficient charge recombination and light emission (see Chapter 3).

Initial studies of organic TFTs and of OLETs were performed on devices using SiO_2/n-doped Si as the gate insulator/gate electrode pair, since this system is well established from amorphous silicon (α-Si) TFTs [7]. The first polymeric insulators (polyimides) for organic thin-film transistors (OTFTs) were used in 1990 [7]. From then, a number of different dielectrics have been explored involving inorganic, organic, and hybrid materials with the goal of high-capacitance, low-leakage films, which are easy to process and compatible with electrode and semiconductor materials.

OTFTs typically operate in accumulation mode [8]. Indeed, for an OSC, when negative or positive gate voltages are applied, holes or electrons are attracted to the gate dielectric–semiconductor interface according to the p-type or n-type nature of the OSC in order to generate an accumulation layer or channel [8–10]. The voltage required to create this channel is largely determined by the capacitance of the dielectric. In the case of a metallic gate electrode, the bias-induced charges are located at the surface of the metal layer. Given the lower polarizability of the OSC layer, the induced charge-carrier density is lower than that of a metal and diffused over a thicker slab of the active layer. This induced charge density distribution at the semiconductor–dielectric interface defines the channel through which current is carried in TFTs.

It was demonstrated by implementing both morphological and optical probe techniques that the charge transport in OTFTs is dominated by the first two monolayers next to the dielectric interface regardless of the type of organic small-molecule semiconductor used [2,3]. In an idealized depiction of the charge accumulation in metal–insulator–semiconductor structure, the charge induced in the semiconductor is equal and opposite to that present at the gate electrode [11]. In real cases, issues of charge trapping in the gate dielectric or at interfaces will modify this simplified picture [6,12].

In the case of using vacuum or air interface between the metal and the semiconductor layers, the charge induced on the semiconductor as a function of gate voltage is $Q = CV = (\varepsilon_o A/t)V$, where ε_o is the permittivity of free space $(8.85 \times 10^{-12}\,F/m)$, A is the area, and t is the distance between the electrode and the semiconductor. In the case where a material is inserted as a dielectric layer, the induced charge for a given voltage is $Q = CV = (k\varepsilon_o A/t)V$, where k is the dielectric constant of the material [11,13].

Increasing the capacitance, C, by increasing the dielectric constant, k, or reducing the film thickness, t, would reduce the voltage necessary to induce the same amount of charge in the semiconductor. For example, in the case of SiO_2 ($k = 3.9$), for a fixed thickness, the amount of charge induced in the semiconductor would be 3.9 greater with respect to vacuum at a fixed bias voltage. This means that a device could operate at ~3.9 lower voltage using an SiO_2 dielectric layer as opposed to vacuum.

It is evident that the optimization of these two parameters, dielectric constant k and thickness t, is essential to selecting and designing an optimal dielectric film.

The dielectric constant of a material is a measure of its polarizability in response to the application of an electric field. This polarizability is the result of reorganization of induced charges, whose timescale determines the frequency dependence of the different contributions (interfacial and space charge, ionic, dipolar, and electronic) to the dielectric constant for a given material.

One example of a dielectric material in which ionic motion is important is barium titanate $BaTiO_3$, which is ferroelectric (i.e., the induced polarization does not decay upon the removal of the electric field). In these structures, titanium ion displacement within its octahedral sites causes extremely large polarizations (2000–3000) [11]. The frequency for ionic motion is up to $10^{12}\,Hz$.

In the case of nonionic materials (such as transition metal oxides), the electronic polarizability is greater for materials containing more electrons (i.e., heavy atoms, greater polarizability). The frequency range for electronic motion is up to $10^{16}\,Hz$.

In the case of the dielectric materials usually implemented in organic thin-film field-effect transistors, electronic and ionic polarization will primarily be responsible for dielectric polarizability.

As we stated before, minimizing the thickness of the dielectric layer is an alternative route for increasing the overall capacitance of the field-effect transistor once the constraints of acceptable leakage currents, effective insulation of the gate, and sufficient dielectric breakdown strength are preserved.

As dielectric films get thinner, the chances for pinhole defects increases, resulting in larger leakage currents. In order to minimize the thickness of the dielectric films, great attention must be devoted to the microscopic and nanoscopic organization of the film. Success has been achieved in generating ultrathin self-assembled monolayer (SAM) films, in which densely packed all-trans aliphatic chains effectively insulate gate dielectrics [14,15]. Typically, the achieved minimum thickness values are on the order of 10–20 nm for inorganic materials such as SiO_2 and 100 nm for polymers, although examples of ultrathin cross-linked polymers have appeared in the literature (vide infra) [16].

For inorganic materials, tunneling currents can also be problematic for ultrathin dielectrics such as SiO_2. Tunneling currents become problematic when band-edge matching between semiconductor and dielectric occurs—for example, in the case of silicon and titanium dioxide [6].

As a thumb rule, the larger the band gap in dielectric materials, the higher the insulation property. However, a trade-off is settled given that the band gap and dielectric constant are often inversely related, so improvement in insulating ability would be offset by lower k.

Thinner films have a higher probability of defects due to the proximity interfaces. Highly defective dielectric films offer conduction pathways via Frenkel–Poole emission or via hopping conduction [6]. Moreover, as films become thinner, dielectric breakdown becomes an issue given that breakdown strength is dependent on the applied electric field, which, as a first-order approximation, equals the applied gate voltage divided by the thickness of the film [13]. As thickness decreases, higher electric fields are observed for a given voltage; for this reason, breakdown can occur at lower voltages.

Finally, dielectric surface roughness is correlated to the reduction of charge-carrier mobility in OSCs due to the disorder induced in the packing structure of OSC thin film. Roughness perturbs the typical π–π stacking in OSC, which is critical to efficient charge transport [4]. Steudel et al. [17] attempted to quantify the impact of surface roughness on mobility for organic TFTs. In this attempt, they used sputtered silica as a dielectric deposited on top of various metal gate electrodes. The roughness of the metal was varied, and the overlying silica gate dielectric replicated this roughness at the dielectric–semiconductor interface. In these studies, it was determined that surface root-mean-square roughness variation, from 1.7 nm for normal thermal SiO_2 to 92 nm for SiO_2 over rough TiW, resulted in a 98% decrease in mobility for pentacene TFTs. The authors propose that the mobility reduction is due to increased grain boundaries in rough films as well as to the irregular channel geometry. Since V_t and detrapping activation energy do not change as a function of roughness, traps are excluded as a possible mechanism for reduced mobility.

Controlling the chemistry and physics of the dielectric–semiconductor interface is essential to control and optimize the performance of organic electronic devices.

There are two main issues associated with the dielectric–semiconductor interface: (i) the electrical properties of this interface, which are essential to the reliability, stability, and threshold voltage; and (ii) the chemical/mechanical properties of the interface, which relate to the ability to organize the semiconductor layer for efficient charge transport. Indeed, in the last 10 years, the major improvements in OLETs optoelectronic performances have been achieved by controlling the OSC conformational and structural organization as a function of the dielectric and morphological features of the underneath insulating layer.

Not only the surface morphology but also the rheological and thermal properties of the polymeric dielectric films play a crucial role in mastering the field-effect charge transport in OTFTs. Indeed, a general and abrupt mobility transition in thin-film semiconductor overlaying polymeric films at temperatures well below the bulk glass transition temperature of the polymers [18] was observed. In particular, when implementing linear conjugated small molecules such as pentacene, the transition from high to low field-effect mobility is closely correlated with large microstructural and morphological alterations of pentacene film growth occurring at the same surface glass transition temperature.

In the case of complete solution-processed active interface, in which the gate dielectric material is deposited from solution onto a solution-processable semiconducting material or vice versa, it is critical to avoid dissolution or swelling effects during deposition of the upper layer: these phenomena can lead to interfacial mixing and increased interface roughness. The cross-linking of the lower layer usually avoid these effects; but also restricts the choice of materials and requires special care to avoid introducing unwanted impurities and trapping groups [19]. The preferential approach is to choose orthogonal solvents for the deposition of the multilayer structure [20]. It has been demonstrated that in this way, solution-processed interfaces can be achieved at which the field-effect mobility is as high as that of the corresponding OSC–SiO_2 interface, for which interfacial mixing is not an issue. This is somewhat surprising since a solution-processed polymer heterointerface is not expected to be sharp and homogeneous: indeed, its width is determined by a balance between entropy favoring a wider interface and the unfavorable energy of interaction between the two polymers [21]. The correlation between interface roughness and mobility also in solution-processed organic field-effect transistors (OFETs) has been investigated by Chua *et al.* [1]. They developed a surface-directing and self-assembling method for spontaneous bilayering two organic bulk phases by solvent evaporation even in the absence of polar substrate interactions. This technique allows the direct formation of a conformal bilayer interface with the fabrication of a robust trap-free and ultrathin gate dielectric. Moreover, they found out that the mobility was constant for low values of the interface roughness of less than a critical roughness threshold. For roughness exceeding this threshold, a very rapid drop of the mobility by orders of magnitude was observed, even for roughness features of surprisingly long wavelength (>100 nm).

The other important issue that can have dramatic effects on interfacial charge transport is the chemical purity and composition of the gate dielectric. The reason for the absence of n-type field-effect conduction in "standard" polymers such as poly(p-phenylene vinylenes) (PPVs) or (P3HT) (poly(3-hexylthiophene-2,5-diyl) with electron affinities around 2.5–3.5 eV has puzzled the community for some time because, in organic light-emitting diode (OLED) devices, many of these polymers support electron conduction.

One of the most revealing studies on the importance of the dielectric semiconductor interface was presented by Chua *et al.* [1]. It was demonstrated that by using appropriate gate dielectrics free of electron-trapping groups, such as hydroxyl, silanol, or carbonyl groups, n-channel field-effect transistor (FET) conduction is in fact a generic property of most conjugated polymers. In contact with trap-free dielectrics such as benzocyclobutene (BCB) based polymers or polyethylenes, electron and hole mobilities were found to be of comparable magnitude in a broad range of polymers. Some polymers, such as poly(2-methoxy-5-(3,7-dimethyloctoxy)-*p*-phenylene-vinylene) (OC_1C_{10}– PPV), even exhibit ambipolar charge transport in suitable device configurations (Figure 4.1).

Indeed, it is possible to achieve electron transport in OFETs by using different polymeric dielectrics including polyethylene, poly(methyl methacrylate), and parylene, although BCB is a preferred choice because of the complete absence of hydroxyl groups. The surface SiOH concentration on standard dry thermal SiO_2 grown at 800–1000 °C has been established to be ~(3–7) 10^{13} cm^{-2} (about 6–12% equivalent of a chemisorbed monolayer). This low concentration is nevertheless electronically significant, as it greatly exceeds the typical carrier concentration of 10^{13} cm^{-2} found in normal FET

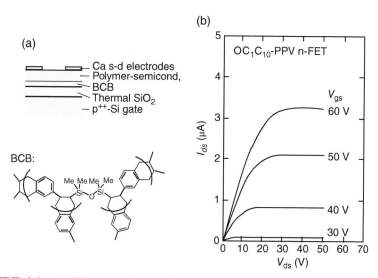

FIGURE 4.1 (a) Diagram of the n-FET and the chemical structure of the cross-linked BCB used as dielectric. Calcium is used for source (s) and drain (d) electrodes. (b) Output characteristics of an OC_1C_{10}-PPV n-FET: $L=200\,\mu m$, $W=10\,cm$. [*Source*: Chua *et al.* [1], Figure 1, p. 195. Adapted with permission of Nature Publishing Group.]

operation. Thus, the mechanism of electrochemical trapping of electrons by SiOH silanol groups is a serious issue in obtaining efficient n-transport in OSCs onto inorganic oxides dielectric.

Before this understanding was developed, it was common knowledge that electrons are far less mobile than holes by a factor of at least 10–100 in OFETs.

Thus, it is clear that whether a particular OSC exhibits n-type or p-type conduction depends more on the device configuration and environmental conditions employed and not merely on the chemical structure of the active materials: when suitable injecting contacts, trapping-free dielectrics, and suitable environmental conditions are provided, most OSCs are capable of electron and hole transport with similar mobilities. It is thus not appropriate to speak of p-type or n-type materials, but one should rather refer to p-channel or n-channel transistors.

The first demonstration of a single-layer and single-material ambipolar OLET has been reported in 2006 based on OC_1C_{10}-PPV [22]. In that case, the dielectric layer was made of a thicker layer of SiO_2 covered by a thin BCB polymeric layer. Despite the fact that SiO_2 is deeply studied and consolidated in inorganic electronics and optoelectronics, an interface BCB layer has been added to SiO_2 to enable effective OLET functioning. The reason of this is strictly related to the role of the dielectric surface chemical composition in generating charge-trapping state.

The implementation of alkyl SAMs for passivating the SiO$_2$ dielectric surface provided not only a further experimental evidence of the electron-transport quenching but also a possible route for allowing the realization of n-type OFETs. Indeed, a systematic study by introducing alkyl-SAMs of various lengths (i.e., hexamethyldisilazane (C1), decyltrichlorosilane (C10), and octadecyltrichlorosilane (C18)) was performed. It was shown that there was no n-FET activity when the SiO$_2$ surface was not passivated. For the alkyl-SAM-passivated SiO$_2$ surface, a short-lived n-channel field effect with the gate threshold shifting to ever higher voltages during device operation was obtained. Longer-chain SAMs suppressed the rate of this shift. However, traps continue to fill and transistor current eventually disappears. These observations indicate that traps originate at or near the SiO$_2$ interface. In general, these traps cannot be completely passivated because siloxane-terminated SAMs cannot completely eliminate surface SiOH groups. The SAM layer, however, provides a tunnel barrier to such sites. When an insulating polymer was used as the buffer dielectric, stable n-channel conduction was observed, as in the case of BCB.

Indeed, the presence of silanol groups also strongly affects the hole transport. By means of a systematic investigation on the role of the dielectric/organic interface in polymer-based OFETs [23], it was found that the higher the silanol-group density on SiO$_2$, the worse the p-type electrical performances. In particular, the authors implemented both superficial chemical treatments such as octadecyltrichlorosilane (OTS) and hexamethyldisilazane (HMDS) and different dielectric polymeric buffer layers such as poly(vinyl alcohol) (PVA), BCB, and poly(methyl methacrylate) (PMMA) for modifying the surface of silicon dioxide dielectric in top-contact and bottom-gate device platform implementing poly(2-methoxy-5-(2-ethylhexyloxy)-1,4-phenylenvinylene) (MEHPPV) and OC$_1$C$_{10}$-PPV conjugated polymers as active layers. It was established that the chemical environment is a predominant factor: in particular, the OH groups present in the PVA chain are less deleterious to hole transport than the silanol groups on the SiO$_2$ surface. Indeed, the best performances have been found when using PMMA, which does not have OH groups in the chemical structure and which shows the lowest affinity with the active polymeric layers.

The functionalization of the OSC/dielectric interface that we have displayed in detail for polymeric semiconductors can be extended to small-molecule OSCs as well. In this case, it is even more evident that the structural and morphological organization of molecular moieties in the first monolayers at the interface with the dielectric layer masters the overall optoelectronic performances of the device.

For demonstrating ambipolar charge transport in single-layer OLET based on small molecules, the well-assessed strategy of SiO$_2$ functionalization was

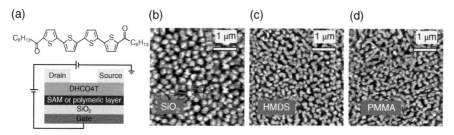

FIGURE 4.2 (a) A schematic of the top contact OLET based on DHCO4T together with the chemical structure of the molecule. (b–d) AFM images of the early growth stages of DHCO4T films sublimed on (b) bare SiO$_2$, and SiO$_2$ treated with (c) HMDS and (d) PMMA. [*Source*: Capelli *et al.* [24]. Adapted with permission of ACS Publications.]

implemented by passivating the dielectric surface by HMDS and OTS SAM treatment and by polymeric buffer film deposition (PMMA and PVA). In Figure 4.2a, the schematic of the first reported ambipolar single-layer OLET device based on α,ω-dihexylcarbonylquaterthiophene (DHCO4T) is shown, together with the molecular structure of the active material [24].

In the case of DHCO4T-based OLET, the different functionalizations of the organic/dielectric interface affect mainly the p-type electrical characteristics, while maintaining unaltered n-type ones. This behavior is exactly the opposite of what was observed for polymer-based OLETs.

Indeed, the origin of the superficial trap states is likely related to the semiconductor film morphology as well as to the chemical interactions with the dielectric surface. In particular, it has been demonstrated that grain boundaries and changes in molecular organization at the microscopic scale can create a distribution of spurious electronic states [24–26]. The existence of energy levels acting as charge traps at the semiconductor–dielectric interface (in the absence of a current flow) increases the gate threshold voltage until their complete filling. Trap states are at the origin, not only of high gate threshold values but also of a fast degradation of the device performance under electrical stress.

Performing a comparative analysis of the morphological features of the first DHCO4T monolayers on top of the different dielectrics by means of atomic force microscopy (AFM) (Figure 4.2b–d), it is possible to observe an initial layer-by-layer growth mechanism with a high density of nucleation centers, independent of the various SiO$_2$ treatments employed. The height distribution of the DHCO4T islands is peaked around a characteristic value of 3 nm. This bidimensional growth affords good film connectivity, essential for a good charge transport. However, a difference in island dimensions can be clearly observed at least with respect to the SiO$_2$ pristine case. The correlation

TABLE 4.1 (a) Electron and Hole Field-Effect Mobility (μ) Data as a Function of the SiO_2 Surface Treatment, Gate Threshold (V_t) Values for Electron and Hole Transport, and the ΔV_t Hysteresis Values.

Dielectric layer	μ_p-Type (cm^2/V s)	μ_n-Type (cm^2/V s)	V_t p-Type (V)	V_t n-Type (V)	ΔV_t p-Type (V)	ΔV_t n-Type (V)
Bare SiO_2	—	—	—	—	—	—
OTS	10^{-6}	10^{-6}	−55	65	>50	>50
HMDS	8×10^{-3}	0.1	−60	53	50	40
PVA	6×10^{-4}	0.5	−22	40	10	15
PMMA	3×10^{-3}	0.15	−46	63	20	18

Source: Capelli *et al.* [24]. Adapted with permission of ACS Publications.

between grains lateral extension and charge trap states allows direct comparison of the morphological features with the OLET optoelectronic characteristics.

Table 4.1 summarizes that the principal limitations to efficient ambipolar transport in DHCO4T-based devices are the large V_t values. In optimum cases, the hole μ is three orders of magnitude lower than the electron μ. This implies that exciton formation is dominated by holes so that the light emission is hole-limited. In spite of this, a μ of ~10^{-3} cm^2/V s should afford a clearly detectable EL signal. On the other hand, high V_t values drastically compress the bias regime where the active channel is open for both electrons and holes. This restricts photon formation to relatively low (absolute value) V_{gs} values and to the saturation condition ($V_{gs} = V_{ds}$). The intermediate gate voltage range, therefore, remains nearly inactive, where ambipolarity regime is supposed to be allowed.

Regarding the role of the chemical interaction between the semiconductor and the dielectric layers in determining the V_t necessary to open the transport channel, the change in the dielectric surface composition principally affects μ, which varies over three orders of magnitude. The differences in V_t as a function of the substrate treatment can be rather large, specifically for the case of PVA.

Thus, this case study highlighted that considering the presence of chemical species at the dielectric–semiconductor interface as the principal cause of V_t variations is an oversimplification in the case of OFETs based on semicrystalline small-molecule active layer.

At the present stage of development of the OLET technology, the use of surface treatments or buffer overlayers is consolidated and widely diffused when the dielectric layer is made of any type of metal oxides and nitrides (SiO_2, Ti_3N_4, ZrO_2, etc.). It is interesting to note that the most efficient OLETs both in single layer [27] and in multilayer [28] architectures are fabricated by implementing only polymer-based dielectric layer (in both cases, PMMA), even though the low intrinsic dielectric constant would make most of the polymeric dielectric not competitive in terms of capacitance.

It is expected for real OLET applications that a high-k dielectric is preferable to a low-k dielectric for a given thickness of dielectric given that the device should exhibit a high drive current at low drive voltage. Various solution-processable high-k dielectrics for low-voltage OFETs have been used in the literature, such as anodized Al_2O_3 [29] ($\varepsilon = 8$–10), TiO_2 [30] ($\varepsilon = 20$–41), or polyvinylphenol loaded with TiO_2 nanoparticles [31]. Many polar, high-k polymer dielectrics, such as polyvinylphenol ($\varepsilon = 4.5$) or cyanoethyl pullulan ($\varepsilon = 12$), are hygroscopic and susceptible to drift of ionic impurities during device operation and thus cannot be used for ordinary thin-film transistor applications [32].

Moreover, SAMs and polymers are also used themselves as new high-k dielectrics. By coating Ta_2O_5 ($\varepsilon = 25$–27) with poly(vinylalcohol), ambipolar transistors with pentacene thin films have been demonstrated [33]. Similarly, coating Al_2O_3 ($\varepsilon = 9.34$) with either Cytop or octadecylphosphonic acid (ODPA), ambipolar transistors with satisfying ambipolar mobility values and relatively low threshold voltage based on naphthalene diimide derivative [34] and inorganic semiconducting nanowire [35] were obtained.

Veres *et al.* [36] have shown that the field-effect mobilities of amorphous poly[bis(4-phenyl)(2,4,6-trimethylphenyl)amine] (PTAA) and other polymers are higher when in contact with low-k dielectrics ($\varepsilon < 3$) than with dielectrics with higher k [37]. The latter usually contains polar functional groups randomly oriented near the active interface, which is believed to increase the energetic disorder at the interface due to the structural disorder in the OSC film, resulting in a lowering of the field-effect mobility (Figure 4.3). Low-k dielectrics also have the advantage of being less susceptible to ionic impurities, which can drift under the influence of the gate field, causing device instabilities.

The authors also showed that the use of low-k dielectrics increases the absolute drain current because the mobility increases more than compensates the charge density decrease that is associated with low-polarity materials.

This apparently counterintuitive result that the implementation of high-k dielectric in (amorphous) polymer-based OTFTs would degrade the electrical performances needed to be further corroborated by systematic variation of the gate dielectrics within the device structure.

Indeed, Naber *et al.* [38] reported the investigation of several gate dielectrics that can be solution-processed on top of F8BT and have a k value below that of PMMA at 3.6, including polyisobutylene, Cytop, poly(isobutyl methacrylate), and polystyrene.

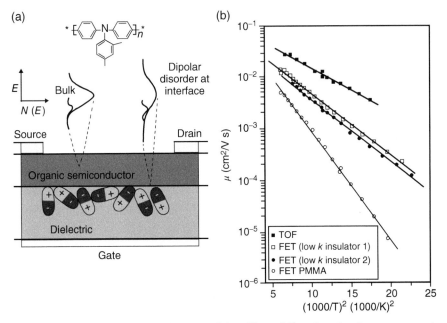

FIGURE 4.3 (a) Schematic diagram of the effect of disordered polar groups on the energetic disorder at the active interface. (b) Temperature dependence of the time-of-flight and field-effect mobility of PTAA. For the field-effect mobility, data for top gate FETs with PMMA gate dielectric and two different lower k dielectrics are shown. [*Source*: Veres *et al.* [36]. Adapted with permission from Wiley.]

The best performance in terms of ambipolar mobilities and current–voltage stability was ultimately obtained with polycyclohexylethylene (PCHE), which has a low k value of 2.3 [39].

The authors proposed that the reasons for the high performance are correlated to the unique combination of properties of the polymer, including the saturated nature of this molecule, a lack of any halogen groups, and a low leakage current.

To enhance the performance further, the high-k poly(vinylidene fluoride-trifluoroethylene) [P(VDF-TrFE)] was placed on top of the PCHE layer that interfaces with the F8BT. The introduction of a material with high dielectric constant (i.e., k value of 50–50 mol% P(VDF-TrFE) is 14, which is six times higher than that of PCHE) aimed at enabling the application of higher electric fields across the PCHE layer without dielectric breakdown, rather than lowering the working voltage bias (and thus the threshold gate voltage). Indeed, P(VDF-TrFE) is well known for its ferroelectric properties [40], but in this bilayer configuration, the occurrence of a ferroelectric polarization is

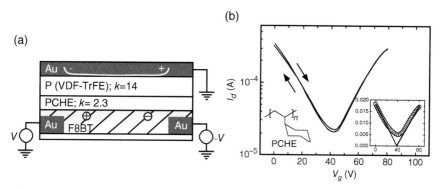

FIGURE 4.4 (a) Illustration of the device scheme with the bilayer dielectric. (b) Transfer measurement at a drain bias of +80V on a OLET with a 120nm thick PCHE and a 260nm thick P(VDF-TrFE) dielectric layer. The dielectric capacitance, channel length, and width were 12 nF/cm², 10μm, and 4cm, respectively. The inset is a plot of the square root of the drain current versus the applied gate voltage. [*Source*: Naber *et al.* [38]. Adapted with permission of AIP Publishing LLC.]

inhibited by the presence of the low-k dielectric because it gives rise to a high depolarization field [41].

With respect to standard PMMA dielectric layer with the same thickness, the achievable hole and electron mobility values are well balanced and maximized (0.01 cm²/V s) (Figure 4.4). Evidently, by interchanging the dielectric that interfaces with F8BT from PMMA to PCHE, the mobility increases by an order of magnitude and the current–voltage hysteresis is also markedly low. Moreover, the achievable current level is 7.7 higher in the case of the dielectric bilayer based on PCHE.

However, the best OLET based on high-k dielectric presented an external quantum efficiency (EQE) comparable to the value that was previously observed for a device with identical geometry and PMMA as dielectric [42]. Thus, the increase in the charge density is not a sufficient condition for increasing the general efficiency in light generation in single-layer ambipolar OLET. Indeed, the EQE initially increases with the applied gate voltage due to the positioning of the emission zone far away from the charge-injecting electrodes; at higher biases, there is a near-linear efficiency decrease by about 25% as the current increases by a factor of 5.

The linear gate voltage dependence together with the magnitude of the efficiency decrease can be correlated to the presence of exciton–charge bimolecular interaction as possible source of efficiency quenching [2,43].

Besides the high electric constant that enables higher accumulated charge carrier density, the intrinsic technological issue related to the use of ferroelectric dielectric such as P(VDF-TrFE) is related to the dipolar polarization of the

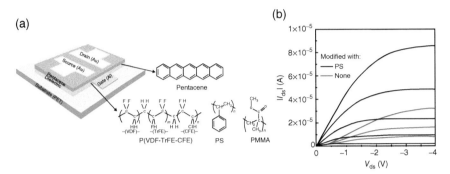

FIGURE 4.5 (a) Schematic diagram of a pentacene-based OFET using P(VDF-TrFE-CFE)/PS and P(VDF-TrFE-CFE)/PMMA bilayers as dielectric. (b) Output characteristics of low-voltage pentacene-based OFET. The gate voltage V_{gs} for different curves of the same device from the top to the bottom are −4.0, −3.5, −3.0, −2.5, −2.0, and −1.5 V, respectively. [*Source*: Jinhua *et al.* [45]. Adapted with permission of Royal Society of Chemistry.]

organic film, which causes large bias hysteresis in the transfer characteristics of the field-effect transistor. A possible strategy is to mix the ferroelectric polymer with other dielectric polymers, such as PMMA, in order to suppress electrical hysteresis. Baeg *et al.* [44] mixed the P(VDF-TrFE) with PMMA to reduce the ferroelectric domains and demonstrated that the ambipolarity of P(NDI2OD-T2) FETs can be tuned.

The application of the high-k terpolymer in OTFTs based on small-molecule OSCs (e.g., pentacene), which normally show higher carrier mobilities than those of conjugated polymers, has been also studied. In these OTFTs, the small-molecule OSCs are deposited by thermal evaporation on the gate dielectrics, using flexible polyethylene terephthalate (PET) as substrate. Therefore, the devices have a bottom-gate top-contact structure. It is notable that the terpolymer is insoluble in many organic solvents (e.g., toluene, isopropanol, anisole, chloroform, chlorobenzene, xylene, water), which allows for the possibility of depositing bilayer with other thin polymer films that are more favorable for charge transport in OSCs as we have already discussed.

The small-molecule based devices are optimized by surface modification of the P(VDF-TrFE-CFE) films with overlaying deposition of different thin polymer films, including polystyrene (PS), PMMA, poly-4-vinylphenol (PVP), and poly(4-vinylphenol-*co*-methyl methacrylate) (PVP-*co*-PMMA) [45]. The optimized OTFTs show carrier mobilities of up to 0.62 cm²/V s at an operating voltage of 4 V (Figure 4.5). In addition, the flexible OTFTs are found to be relatively stable even after 1000 bending tests under a tensile strain of about 1%. Therefore, the P(VDF-TrFE-CFE) terpolymer is a promising gate dielectric material for low voltage, flexible, and large-area organic electronic devices and circuits based on small-molecule OSCs.

Besides ferroelectric dielectrics, there are several other high-k polymer–polymer blend and inorganic–polymer blend dielectrics that have the potential for inducing high carrier density in ambipolar transistors, for example, sodium beta alumina [46], self-assembling nanodielectrics (SANDs) [47], and cross-linked inorganic/organic hybrid blends [48]. These dielectrics have demonstrated to be effectively applied to both p-type and n-type unipolar transistors. However, to date, their use on semiconducting materials capable of ambipolar transport has never been shown.

As alternatives to the conventional dielectrics and even to high-k dielectrics, electrolyte dielectrics can be implemented in field-effect device architecture given that very high carrier density can be accumulated at their interface. Electrolytes are ionic conductors, but electron and hole insulators. Therefore, the prompt motion of the positive and the negative ions under an electric field can create strong accumulation of space charges at the electrolyte/semiconductor interface, in the form of an electric double layer, known as Helmholtz layer [49,50] (Figure 4.6).

This electric double layer acts as a nanogap capacitor with huge capacitance and electric field; the capacitance is determined by the shortest distance of the nearest ionic molecules to the ionic-liquid/semiconductor interface (≈ 1 nm). As a result, a very high concentration ($\approx 10^{14}$ cm^{-2}) of free sheet carriers (electrons or holes) inside the semiconductor can be induced electrostatically.

Therefore, electric double-layer gating is able to accumulate more charge carriers by applying only very low gate voltage in comparison with conventional solid–dielectric gate (i.e., SiO_2).

In particular, a special class of solid polymer electrolytes known as ion gels [51] can serve as high-capacitance gate dielectrics in OTFTs operating at much higher frequencies, currently up to 10 kHz. The faster polarization response is a manifestation of both very large concentration and mobility of ionic species in the gels. In a milestone work, Cho *et al.* [52] also demonstrated that ion gels are readily printable on plastic substrates by using a commercial aerosol jet printing technique to print all components (i.e., the metal electrodes, polymer semiconductor, and dielectric) of ion-gel gated OTFTs on plastic. Ion gels have several very attractive characteristics for printed electronics, the most intriguing of which is that their high polarizability allows the gate electrode in the field-effect transistor to be physically offset from the channel. This feature enables simpler device architectures, in which the usual requirement for precision registration of the gate over (or under) the source–drain channel is relaxed, significantly enhancing the printability of

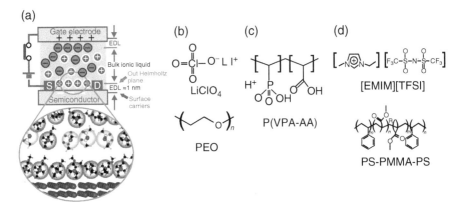

FIGURE 4.6 (a) Schematic of charge accumulation by an electric double-layer (EDL) formed at the interface between a liquid (ionic conductor: ionic liquid) and the solid (electron conductor: semiconductor). Cations (red circles) and anions (green circles) can be electrostatically accumulated onto the channel surface by applying a positive and negative gate voltage, respectively. The inset shows magnified schematics of the interface between the semiconductor and the ionic liquid. Different types of electrolyte gating materials: (b) dissolved ionic species in a neutral polymer, for example, LiClO$_4$ in polyethylene oxide; (c) polymer electrolyte where the polymer itself is charged, for example, P(VPA-AA); (d) ionic liquid, which is a molten salt such as [EMIM][TFSI] that can be used as it is included into a block copolymer matrix (e.g., PS-PMMA-PS) to form a ion gel. [*Source*: Bisri *et al.* [50]. Adapted with permission from Wiley.]

the devices. Furthermore, devices based on ion-gel dielectric have good ambient and operational stability, as well as very large transconductance (>500 μS/mm), which may make them useful for applications where significant drive currents are required.

In spite of the excellent results reported for OTFTs, an extensive implementation of high-k dielectrics in OLETs has not yet been reported. OLET technology development indeed is still strongly related to fundamental scientific issues concerning new light-emitting material synthesis and device architecture design. A straightforward technological transfer of the dielectric fabrication know-how between OTFT and OLET technology is expected.

Nevertheless, dielectrics specifically implemented in OLETs should satisfy further constraints in order to fully exploit their potential in organic photonic applications. In particular, multifunctional dielectrics capable of light handling are technologically relevant for enhancing optoelectronic performances in OLETs. For example, optically transparent dielectric and gate electrode onto glass substrate are the preferential platform for OLET realization in order to avoid unwanted light absorption processes. Indeed, transparent and high-k dielectrics are required for modulating light confinement and waveguiding within the device multilayer structure. The control of light outcoupling is a mandatory engineering step for allowing the realization of highly integrable organic optoelectronic devices for low-loss light-signal transmission in photonic circuitry.

Moreover, light waveguiding and reflectivity properties of the dielectric may enhance the OLET brightness and efficiency by harvesting the emitted photons in preferential direction according to suitable light emission spatial profiles. The realization of transparent dielectric layers is straightforward in most of the cases due to the limited dielectric film thickness and to the intrinsic electronic properties required for an insulating material, that is, high valence to conduction band gap. This condition favors both the film insulation properties and the film transparency in the visible range. On the contrary, in order to use the dielectric as reflector, or waveguide, or lasing resonant cavity, a more complex design approach is to be assessed.

4.2 EMISSIVE AMBIPOLAR SEMICONDUCTORS

OLETs require the active materials to possess both high carrier mobility and high photoluminescence quantum yields (PLQYs) to realize the potential applications in active-matrix full color display, integrated photonic circuitry, sensing, and possibly the long-searched electrically driven organic laser [53]. However, in most cases, OSCs with high carrier mobility due to the strong intermolecular π–π stacking suffer from luminescence quenching in the solid state in correlation to the singlet fission or exciton quenching. For example, rubrene, tetracene, and pentacene exhibited superior FET mobility, but they exhibit low or no emission [54,55].

There are several reasons for luminescence quenching in the solid state, which include fission of a singlet exciton to give two nonemissive triplets [54], as well as exciton quenching on the defect sites. On the other hand, some conjugated molecules, for example, tetraarylethylenes [56], show enhancement of photoluminescence in the solid state (aggregation-enhanced emission) [57]. One of the simplest blue emitters, anthracene, has a PLQY of 64% in crystals [58], but no thin-film transistors could be so far produced from this material. The larger aromatic core in tetracene gives rise to reasonable charge mobility of approximately $0.1\,cm^2/V\,s$ in thin-film FETs [59], but the solid-state PLQY drops down to 0.8%, mostly as a result of singlet fission [60]. The tetraphenyl derivative of tetracene, rubrene, has a large FET mobility of up to $10–20\,cm^2/V\,s$ in single crystals [61], but a correspondingly very low (less than 1%) PLQY in the solid state, which is also attributed to singlet fission [53]. Pentacene, a benchmark p-type OSC, has a mobility above $1.0\,cm^2/V\,s$ [62] and does not fluoresce in the solid state. A number of other high-mobility OSCs have been reported [63], including dinaphthothienothiophene and its derivatives, with a mobility value of $8\,cm^2/V\,s$ or less in thin films [64] and $16\,cm^2/V\,s$ in single crystals [65]. However, these compounds all have low or no emission in the solid state.

The overall picture emerging from the literature is that maximizing the charge mobility in OSCs (both in thin films and single crystals) necessarily leads to decreased emission efficiency [57].

As it will be discussed in Chapter 5, the charge-transport mechanism in the OSCs, which are widely implemented in field-effect transistors (such as polycrystalline thin films of small molecules and microcrystalline polymers) is described as a middle case between the two extremes of thermally activated charge-hopping model typical of disordered semiconductors and the band theory typically used to rationalize the long-range-order single crystals. In organic systems, the electronic coupling of adjacent molecules and the polaronic relaxation energy, which is the energy gained when a charge geometrically relaxes over a single molecule or polymer segment, is an important parameter determining the probability of charge transport from one molecule to another and depends strongly on the particular molecule and the relative position of the interacting units.

OSC systems that display high charge mobility in field-effect device structures usually tend to show strong interchain/intermolecule interactions (especially for low-molecular-weight oligomers). This feature in turn can be considered as a drawback in case these materials are to be implemented in optoelectronic devices. Indeed, rigid oligomers present a relatively low luminescence efficiency in the solid state, much lower than that of polymers such as polyphenylene vinylenes and polyfluorenes [66]. Whereas in solution, the PLQY of poly(3-alkylthiophenenes) is 30–40%, it drastically drops to 1–4% in the solid state due to increased contribution of nonradiative decay via interchain interactions and intersystem crossing caused by the effect of the sulfur heavy atom [67]. For the same reason, thiophene-based polymers have a stronger spin–orbital coupling than benzene-based polymers; hence, triplet-state processes play a greater role in their photophysics [68].

The fact that strong π–π interactions that are required for high charge mobility necessarily lead to quenching of the luminescence in the solid state is considered as a thumb rule in choosing the proper organic conjugated semiconductors (being small molecules or polymers) to be used in single-layer single-material ambipolar OLETs. In real ambipolar OLETs, the number of emitted photons is dependent on the flowing charge density and on the rate of formation of emissive excitons: significant effort is put by the scientific community to synthesize and/or process active materials capable of balancing these two competing processes.

Nevertheless, most recently, the maximized charge mobility and PLQY have been achieved by Perepichka and coworkers [69] using 2-(4-hexylphenylvinyl) anthracene (HPVAnt) as an OLET material in the crystalline state. Another possible candidate for OLET materials is thiophene/phenylene co-oligomers, which possess high luminescence efficiency and ambipolar FET behavior [70].

Indeed, it has been demonstrated that the structurally simple hydrocarbon HPVAnt exhibits a unique combination of high field-effect mobility of up to $1.5\,cm^2/Vs$ in thin films ($2.6\,cm^2/Vs$ in single crystals), strong solid-state emission, and excellent operational stability in air. Both thin-film and single-crystal transistors emit bright-blue EL that is stable under continuous operation. Moreover, the high PLQY of 70% that was achieved for HPVAnt crystals is likely attributable to the high energy of the triplet excited state T_1 relative to singlet S_1 ($2T_1 > S_1$), which turns off emission quenching through singlet fission and, thus, outlines a new principle for the design of emissive crystalline OSCs [70].

Molecular design and implementation of OSCs are widely believed to be the most efficient routes to improving the device characteristics of OFETs. However, previous molecular design efforts were generally focused on efficient carrier transport rather than on enhancing multifunctional characteristics such as light emission and light sensing. Thus far, most reported multifunctional materials were discovered by serendipity, and a deep understanding of the relationship between the molecular structure and the multifunctional properties remains a major challenge.

Application of well-developed molecular design strategies can improve the exploration efficiency of light-emitting semiconductors. According to systematic studies on typical OFETs, the properties of OSCs can be modulated by fine-tuning the following parameters: (i) the energy level, to facilitate carrier injection from contacting electrodes into semiconductors, to modulate carrier trapping at the dielectric semiconductor interface, and to influence exciton separation as well as carrier recombination in the conductive channel; and (ii) intermolecular overlap, packing, and interaction, to facilitate charge transport. Fortunately, many molecular design strategies have been developed in the recent years to modulate these properties, including, in particular, light emission. Particularly, the introduction of specific functional groups in the existing carrier-transport materials is a shortcut to take advantage of these strategies and to obtain novel multifunctional materials [71].

For instance, a possible strategy for achieving ambipolarity in novel OSCs is to synthesize small molecules or copolymers with electron-donor moiety (D) and electron acceptor (A). The interchain D–A interaction strengthens intermolecular interaction that reduces the π–π stacking distance, increasing the self-assembly capability. In principle, by locating the charge-transfer process onto the most emissive moiety or introducing a further moiety devoted to excitons formation and efficient radiative deactivation, it is possible to engineer the functionalities (hole/electron charge transport and light emission) within the single molecular or polymeric unit.

Among the different acceptor units, naphthalene diimide (NDI) derivatives and diketopyrrolopyrrole (DPP) derivatives are the most promising in achieving high-mobility ambipolar transport. The electron-accepting NDI unit has strong π–π interactions that provide better charge-transporting properties between chains, and the diimide group pulls down energy in the lowest unoccupied molecular orbital (LUMO) level to make the n-type transport air-stable. Copolymers of dialkyloxydithiophene and NDI result in ambipolar charge transport in transistors, with electron mobility as high as $0.02\,cm^2/V\,s$ and hole mobility still in the range of $10^{-3}\,cm^2/V\,s$ [71]. By covalently connecting benzothiadiazole units through thiophene linkers to the NDI unit demonstrated significant improvements of the balance of the carrier mobility values ($0.05\,cm^2\,V/s$ for electron and $0.1\,cm^2/V\,s$ for hole) with good air stability [72]. Recently, Lee *et al.* [73] reported very high ambipolar mobility from a solution-processable D–A copolymers consisting of DPP acceptor with siloxane-solubilizing groups and selenophene donor, named PTDPPSe-Si. Mobilities as high as 3.97 and $2.20\,cm^2/V\,s$ for holes and electrons, respectively, were achieved. Modifying the alkyl side chain is known to induce denser molecular packing. PTDPPSe-Si modified with pentyl chain demonstrated ambipolar transport with unprecedented high hole and electron mobilities of 8.84 and $4.34\,cm^2/V\,s$, respectively, which are the highest values for ambipolar polymer FET [74].

Despite the complex task in engineering effective multifunctional organic conjugated semiconductors, it has already been demonstrated that the minimized excitonic losses in ambipolar OLET allowed higher EQE stability at higher current density with respect to OLEDs. For example, Zaumseil *et al.* [22] demonstrated that ambipolar OLET based on conjugated polymer can achieve ambipolar current density up to $50\,A/cm^2$, which is comparable to what was obtained with rubrene single-crystal ambipolar OLET. EQE stability is observed in device operating at current densities as high as $4\,kA/cm^2$ [75].

This is in sharp contrast with the values achieved by conventional OLEDs, where the EQE starts to suffer a catastrophic roll-off at a current density as low as $1\,A/cm^2$ [76] (see Chapter 2).

The direct comparison of the EQE of the ambipolar OLET with the one of OLED remains a great challenge. Despite intense electroluminescence being obtained not only from ambipolar OLETs based on small molecules [77–80] but also from polymers [22,38,81] as well as organic single crystals [46,67],

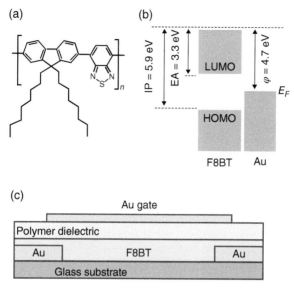

FIGURE 4.7 (a) Chemical structure of F8BT. (b) Highest occupied molecular orbital (HOMO) and lowest unoccupied molecular orbital (LUMO) levels of F8BT in relation to the work function of gold; IP: ionization potential, EA: electron affinity, E_F: Fermi energy level, ϕ: work function. (c) Schematic illustration of a bottom contact/top gate polymer transistor with gold source/drain electrodes, F8BT as the semiconducting and emissive polymer, a spin-cast insulating polymer as the gate dielectric, and an evaporated top gate electrode. [*Source*: Zaumseil *et al.* [22], Figure 1, p 70. Adapted with permission of Nature Publishing Group.]

most of those ambipolar OLETs were fabricated with a single OSC that acts as both the transport and the emitting layer. This is in contrast to OLEDs, which usually use specialized layers where electron transport, hole transport, and emission are each optimized in a distinct layer.

Among the plethora of light-emitting conjugated polymers synthesized in the last 20 years, poly(9,9-di-*n*-octylfluorene-*alt*-benzothiadiazole (F8BT) (Figure 4.7a) [22] showed the most performing results for real applications in polymeric single-layer OLETs. F8BT is an efficient green-light emitter with photoluminescence efficiencies of 50 to 60% in solid films [82]. Because of the electron-withdrawing properties of the benzothiadiazole group (BT), F8BT exhibits a relatively high electron affinity (EA) of 3.3 eV and a large ionization potential (IP) of 5.9 eV (Figure 4.7b). It is often used as an electron transporter in polymer-blend OLEDs [83] and photovoltaic cells [84].

As we have already discussed in the previous paragraph, PMMA is selected as dielectric because it is known for its good dielectric characteristics (dielectric constant $\varepsilon = 3.6$), is soluble in solvents orthogonal to F8BT, and

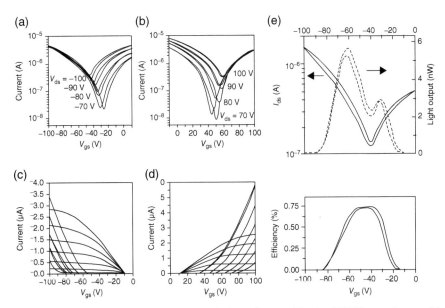

FIGURE 4.8   (a, b) Transfer characteristics of an ambipolar F8BT transistor with PMMA as the gate dielectric (capacitance $C_i = 4.2\,nF/cm^2$, length $L = 20\,\mu m$, width-to-length ratio $(W/L) = 500$) at different negative and positive V_{ds}. (c, d) Output characteristics of the same transistor for different V_{gs} (from 0 to $-100\,V$ and from 0 to $100\,V$ in steps of $10\,V$). Saturation mobilities of holes and electrons for this device were $7.5 \times 10^{-4}\,cm^2/V\,s$ and $8.5 \times 10^{-4}\,cm^2/V\,s$, respectively. (e) Transfer characteristics, corresponding light output, and external quantum efficiency of a F8BT in the same device geometry as reported in the previous characteristics. [*Source*: Zaumseil *et al.* [22], Figure 4, p. 73. Adapted with permission of Nature Publishing Group.]

contains no or a small number of —OH groups that could potentially trap electrons. The fabricated transistors showed very clean ambipolar transport with nearly perfectly matched, gate-voltage-independent saturation mobilities of $7-9 \times 10^{-4}\,cm^2/V\,s$ for electrons and holes.

Turn-on voltages, extracted from transfer characteristic at low source–drain voltages, are about $30\,V$ for electrons and $-20\,V$ for holes. The output characteristics show good saturation as expected for devices with a channel length significantly greater than the gate dielectric thickness (Figure 4.8a–d).

Figure 4.8e shows the transfer and corresponding light output characteristics of a F8BT OLET with a channel length of $75\,\mu m$ and width of $4\,mm$.

Two peaks in the light intensity are observed when either hole or electron currents dominate, and a local minimum is evident when the drain current reaches a minimum. Referring to the EQE of emission in Figure 4.8e, however, a global maximum plateau around the drain-current minimum is found.

This corresponds to the gate voltage region where the emission zone is positioned within the channel, as expected.

The main drawback of this configuration is related to the use of same-material (gold) electrode for injecting both electrons and holes, with plausible presence of contact resistance in the case of electron injection. Even though charge injection is not supposed to occur only at the edge of the electrodes, but rather over a larger area away from the edge given the overlap of the gate electrode and the source and drain electrodes, improvements in charge-injection process is required as it will be described in the next paragraph. Indeed, improvements in device performances are obtained by functionalizing the source and drain contacts with different SAMs [84].

Recently, the further improved and optimized device structure using zinc oxide as electron-injecting layer and reflective gate electrode allowed to reach very high brightness values, up to $8000\,cd/m^2$, and EQE up to 8% without varying the active material or altering the process protocol used [27].

It was observed that the long-range crystallite organization in F8BT thin film can account for variation into charge-carrier mobility and electroluminescence emission zone location in the channel. Crystallization in semicrystalline polymers leads to chain orientation variations on mesoscopic length scales and to grain boundaries with amorphous interstices. Such spatial differences in microstructure are expected to affect local charge-carrier mobility and current density [85], exciton formation, diffusion, and quenching [86].

Zaumseil *et al.* [87] used this feature to correlate polymeric semiconductor microstructures (amorphous, polycrystalline, and aligned nematic) with charge transport paths in ambipolar OLETs.

Particularly, F8BT films obtained from anhydrous xylene solution of polymer with number average molar mass of $M_n = 62\,kg/mol$ are subjected to different deposition treatments: (i) annealing at 290 °C for 30 min and slowly cooled to induce polycrystallinity with large crystallites [82]; (ii) annealing at 120 °C (below glass transition temperature) for obtaining amorphous films without any microstructural features; and (iii) deposition onto a polyimide layer, which was spin-coated onto the glass substrate, cured in nitrogen, and mechanically rubbed before source/drain gold electrodes deposition. The latter treatment allowed to obtain a nematic-phase F8BT film with polymer chains predominantly aligned along the rubbing direction after annealing up to F8BT liquid crystalline melt and quenching.

Optical imaging of the emission zone is performed on OLETs based on active layers subject to different process protocols, which lead to polycrystalline and amorphous F8BT (Figure 4.9a and b) [87].

While light emission from the amorphous F8BT film is quite narrow (width at half maximum of 2 μm) and featureless, the emission zone of the

(a)   (b)

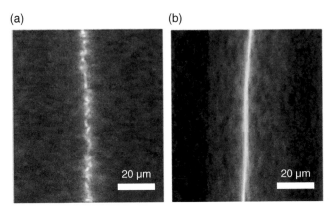

FIGURE 4.9   Optical image of emission zone of (a) polycrystalline and of (b) amorphous illuminated F8BT-OLETs in ambipolar bias conditions. [*Source*: Zaumseil *et al.* [87]. Adapted with permission of AIP Publishing LLC.]

polycrystalline polymer film is much broader (4 μm) and shows bright spots of various shapes and sizes and dark gaps. Although the intensity distribution in Figure 4.9b immediately appears to be related to the polycrystalline morphology of the F8BT film, it is unclear whether this is due to variations of photoluminescence efficiency, outcoupling efficiency, current density, or variations in nonradiative recombination due to trap sites.

The authors collected EL optical images during a gate voltage sweep at constant V_{ds}: the overlay of the recorded images allowed to generate a map of electroluminescence of the whole channel region.

This EL map showed that certain areas of the channel region never lit up and remained dark under all voltage conditions, while the bright regions of the channel formed a type of fractal network. The bright areas were isotropic and the brightness histogram showed one broad distribution and a second narrow distribution at higher brightness, indicating strongly preferred areas of emission.

These variations of the EL emission distribution within the channel cannot be ascribed to differences of PL or outcoupling efficiency. Moreover, extended, long-lived charge-trapping and quenching sites can be ruled out as the origin of dark areas because the EL maps for forward and reverse voltage sweeps, that is, from electron to hole accumulation and vice versa, are identical and do not show any hysteretic features. According to the authors, this evidence strongly suggests that lateral variations in current density are responsible for the observed variations in emitted light intensity. In a polycrystalline film with domains of different orientations, charge carriers are more likely to pass through F8BT regions aligned along the lateral field,

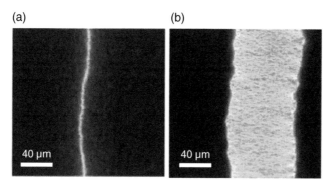

FIGURE 4.10   Optical image of the emission zone (a) and accumulative EL image (b) of the OLET with partially aligned F8BT collected during transfer characteristic. [*Source*: Zaumseil *et al.* [87]. Adapted with permission of AIP Publishing LLC.]

circumventing those aligned perpendicularly. However, the EL intensity is not proportionally linked to the carrier mobility but instead reflects in a complex manner the lateral variations in mobility, polymer orientation, and electric field along the current conduction paths for both holes and electrons between source and drain. Indeed, in an amorphous polymer, the carrier mobility is isotropic and should hence lead to a uniform current density and a featureless emission zone, as seen in Figure 4.9b.

However, when the F8BT film is thermotropically aligned on a rubbed polyimide layer (Figure 4.10a), the current density and emission pattern are evidently influenced by polymer chain alignment and, thus, distribution of high and low mobility areas in the device. Indeed, hole and electron mobilities are significantly higher than those of a polycrystalline F8BT film.

In the EL map (Figure 4.10b), a strong anisotropy of the bright areas and thus current paths is revealed: clearly preferred current paths are present and the conduction appears to be of filamentary nature. Due to the almost complete alignment, charges rarely need to circumvent low mobility areas and thus less bright-dark contrast occurs. Finally, the intensity is higher and its distribution is narrower than in the case of polycrystalline F8BT.

It is evident that substantial modifications to the device processing have to be made to achieve the simultaneous optimization of the optical and electronic properties in single-layer ambipolar OLETs. Particularly, this step is technologically mandatory for fine-tuning the photonic characteristics of the transistor-based devices.

Optical waveguiding and gain/loss properties in organic conjugated polymers are deeply affected by postdeposition annealing process after the spin-coating [88], which in turn modulates the optoelectronic performances of the polymer once used as active material in field-effect transistors.

In the case of F8BT ambipolar OLET, it was observed that the best transistor performance is achieved by annealing the polymer at a temperature above the liquid crystalline melting temperature: however, this postdeposition annealing process is not compatible with waveguiding and lasing applications.

Moreover, the optimization of the optical characteristics, such as amplified spontaneous emission (ASE), which represents a good figure of merit for efficient waveguiding and gain, imposes further constraints on the active-layer thickness. Indeed, the device active layer has to be increased up to 70–75 nm in order to (i) exceed the waveguide cutoff thickness, (ii) increase the mode confinement within the polymer semiconductor [89], and (iii) improve the isolation of the mode confined in the polymer semiconductor from the absorbing gate electrode. The latter is separated from the polymer semiconductor by a PMMA buffer layer that cannot be made too thick because it needs to act as an effective gate dielectric.

This condition in device structure may be detrimental for the optoelectronic performance given that in the top gate/bottom contact configuration charges need to be transported from the electrodes through the bulk of the film, and for such thick films, contact resistance, particularly for electron injection into the channel, becomes significant [90,91].

Gwinner et al. [92] found that annealing F8BT layer at high temperatures, ideally at 290 °C, significantly improves the ambipolar transistor performance by reducing threshold voltages and hysteresis. This is particularly important for strong light emission since a wider voltage range for ambipolar operation is possible. Figure 4.11a shows a comparison of the transistor source–drain currents I_{ds} during a transfer scan. The transistors with F8BT processed at 290 °C show the best performance, whereas reduced annealing temperatures yield lower transistor currents for both electrons and holes. In the ambipolar operation regime, the current is already more than halved for 240 °C. It further drops to 15% and only 4% for annealing temperatures of 160 and 120 °C, respectively. Moreover, the plot of the corresponding photocurrents (I_{ph} in Figure 4.11b) shows that the maximum light intensity drops approximately linearly by two orders of magnitude between the two extreme annealing temperatures.

The decrease of the electron mobility from 120 to 160 °C can be explained by conformational changes of the F8BT chains stacking below and above the glass transition temperature. Between 120 and 160 °C, a rearrangement of the F8BT chains occurs into an energetically more favorable packing structure, which exhibits less overlap between benzothiadiazole units of neighboring polymer chains, making the electron hopping transport between those sites less favorable [82].

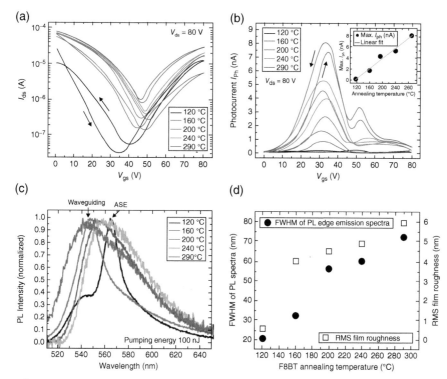

FIGURE 4.11 (a) Comparison of the OLET ($L = 10\,\mu m$, $W = 4\,cm$, 70 nm F8BT, 325 nm PMMA, 15 nm Au gate) transfer characteristics for F8BT annealing temperatures between 120 and 290 °C. (b) Corresponding photocurrents during the transfer sweeps. The inset shows the respective maximum photocurrents. (c) Comparison of the PL spectra at the sample edges of 155 nm thick F8BT films annealed at temperatures between 120 and 290 °C while being excited with an energy of 100 nJ. (d) Corresponding FWHM of the spectra and RMS film roughness. [*Source*: Gwinner *et al.* [92]. Adapted with permission of ACS Publications.]

On the other hand, only the F8BT films annealed at 120 °C showed clear evidence of ASE process to take place with the photoluminescence (PL) spectrum collapsing into an ASE peak centered at 563 nm (Figure 4.11c), which is a clear evidence of efficient light waveguiding in the polymeric thin film [93,94].

For higher annealing temperatures, however, ASE behavior cannot be observed, even for higher pump energies. Instead, the spectra only show a slightly modified PL of F8BT with peak around 550 nm and decreasing intensity for higher temperatures. Decent waveguiding can still be obtained for 160 °C, although the peak full-width-at-half-maximum (FWHM) value of 32 nm is likely to overestimate the waveguiding ability as the spectrum does not show ASE. For higher temperatures, the FWHM steadily increases and intensity decreases, indicating major losses and poor waveguiding.

A phenomenological explanation for the lack of trade-off for electronic–optoelectronic and photonic features is found in the observation that increasing the annealing temperature above 200–400 °C induces progressing crystallization of the active layer with consequent increase of grain size. The crystallization process is attributed to the polymer entering into a liquid crystalline mesophase during the annealing with much increased chain mobility: this process allows from one side the increase in the charge-carrier mobilities (particularly, electron mobility) with consequent enhancement of ambipolarity and light-emission intensity, but from the other side, the suppression of efficient light waveguiding and optical gain within the active layer given the severe light scattering at the grain boundaries.

In summary, the OLET processing and, in particular, the effects of post-spin-coating annealing modulating the morphology of an F8BT film, are of crucial importance for efficient light-waveguiding and amplifying applications.

It was proposed to functionalize the charge-injecting electrodes with different types of SAMs for partial recovery of the device electrical performance when the device is configured for maximizing the photonic features of the active material. Indeed, a reduction of the high contact resistance is expected when tuning the charge-injection process.

However, this procedure only slightly affects the arrangement and conformation of the polymer moieties at the nanoscale, which are the main causes of self-excluding optimization of either the electrical or photonic performances of the device in the bottom contact/top gate configuration.

Other methods can be implemented for improving ambipolar injection and charge transport within the polymeric active material without introducing modifications of the device structure in terms of interface functionalizations, injecting electrodes, and gate dielectric.

Even though in the bottom contact/top gate configuration the overlap of the gate electrode with the source–drain electrodes support injection of charges even when a high injection barrier is present, all charge carriers have to travel through the undoped polymer film of low conductivity to reach the interface with the dielectric and enter the charge accumulation layer. For this reason, the contact resistance increases with film thickness.

As we will discuss in the next paragraph, a general strategy to reduce the high contact resistance, regardless of the device configuration, is to use two different metals: a high-work-function metal for hole injection (e.g., Au, Pd) and a low-work-function metal (e.g., Ca, Mg, Al) for electron injection, similarly to what is done in OLEDs. However, low-work-function metals are not air-stable, and patterning is typically limited by shadow masking.

A simple method that significantly improves the injection of both holes and electrons into large-band-gap polyfluorenes such as F8BT in a bottom contact/top gate FET structure consists in adding small amounts of single-walled carbon nanotubes (SWNTs) to the semiconducting polymer without creating percolation paths between the source–drain electrodes. Carbon nanotubes have been of interest for charge injection in organic electronics for some time. Several groups have reported examples of injecting electrodes for OFETs where the main electrode material was a metal or conducting polymer and carbon nanotubes extended from the edge of the electrode laterally into the channel [95,96].

Among the many techniques [97,98] developed in the past several years to select semiconducting SWNTs from a mixture, one of the most interesting is the polymer-assisted separation method, because of its high effectiveness and scalability [99]. Recently, highly enriched selection of small-diameter semiconducting nanotubes using a polyfluorene derivative (PF8) [100] was reported. Later, it was demonstrated that nanotubes with various diameters, including the ones with large diameters (ca. 1.5 nm), can be selected by exploiting polyfluorene derivatives bearing side chains of different lengths [101].

Zaumseil *et al.* [102] investigated the influence of small amounts of semiconducting SWNTs dispersed in polyfluorenes such as F8BT on device characteristics of bottom contact/top gate ambipolar light-emitting field-effect transistors. The authors found that the presence of SWNTs within the semiconducting layer at concentrations below the percolation limit significantly increases both hole and electron injection without leading to significant luminescence quenching of the conjugated polymer. As a result of the reduced contact resistance and lower threshold voltages, larger ambipolar currents and thus brighter light emission are observed. Energy transfer from the polymer to the nanotubes, which emit near-infrared light [103], is a possible drawback of this method. In practice, the limited concentration of nanotubes does not quench the visible F8BT emission significantly.

This method does not require any additional patterning steps and is compatible with high- and low-temperature processing. It leads to lower contact resistance and thus lower threshold/onset voltages for both holes and electrons, higher ambipolar currents, and emission intensities. The observed reduction in onset voltages for electron and hole transport is at least partially due to lower contact resistance. The source–drain currents at low source–drain voltages in the output characteristics, plotted in Figure 4.12a and b, are significantly larger in devices with SWNTs than in those without, indicating substantially lower, yet still nonohmic, contact resistance. The EQE measured with a silicon photodiode in the forward direction through the semi-transparent gate electrode and plotted in Figure 4.12c is similar to the best

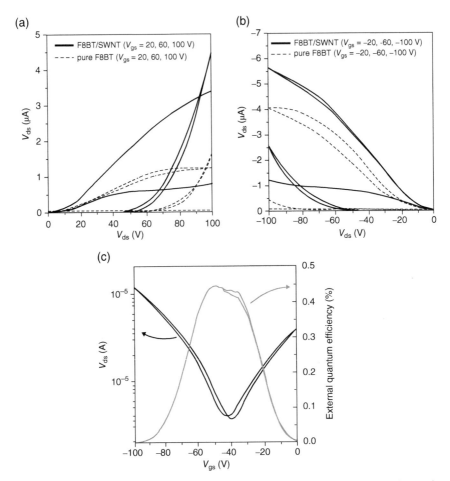

FIGURE 4.12 Comparison of output characteristics (a, b) at positive and negative gate voltages of OLETs with pure F8BT (dotted lines) and F8BT/SWNT (solid lines) films ($L = 20\,\mu m$, $W/L = 1000$, $C_i = 4.4\,nF/cm^2$). (c) Current–voltage characteristics and external quantum efficiency of green light emission (maximum at 560 nm) from OLET with F8BT doped with SWNT thin films ($V_{ds} = -100\,V$, $L = 20\,\mu m$, $W/L = 1000$, $C_i = 4.4\,nF/cm^2$). [*Source*: Gwinner *et al.* [102]. Adapted with permission of ACS Publications.]

pure-F8BT devices. Note that using this method, the obtained EQE is under-estimated, as not all of the emitted photons but just the small fraction transmitted through the gold gate are collected [104].

The observed ambipolar injection improvement is applicable to most conjugated polymers in staggered transistor configurations or similar organic electronic devices where injection barriers are an issue.

In general, the excellent optical and photonic characteristics together with the process versatility and scalability are supposed to expedite the implementation of light-emitting ambipolar conjugated polymers in the realization of OLETs for real applications. However, as we discussed in detail, the charge-transport performances of the most promising candidates reported in the literature are good in terms of ambipolarity balance, but limited with respect to the maximum achievable field-effect mobility.

The use of effective multifunctional OSCs endowed with efficient hole and electron field-effect charge transport capability and light-emission functionality is the real enabling step in the realization of single-layer and single-material ambipolar OLETs.

The condensed structure of the organic active layer, with tight intermolecular packing and flat morphology, can play an important role in determining the device performance and functionality of OLETs. In many cases, organic crystals and thin films have two or more structural phases upon different environments, and these different phases possess different intermolecular packing and charge transport properties. Transition between these phases can be achieved by controlling the temperature, pressure and even the surface energy of gate dielectrics.

Morphology is an important factor of a condensed structure. It is well-known that the morphology of an organic thin film is strongly related to the device fabrication techniques and procedures. For example, the deposition rate and substrate temperature significantly affect the morphology of vacuum sublimed films, while solution concentration, spin rate, and annealing temperature have a significant impact on the morphology of spin-coated films.

The use of linear conjugated oligomers as active materials in field-effect organic transistor guarantees, in principle, higher electrical performances with respect to polymers, thanks to the long-range order molecular organization achievable in thermally evaporated thin films. Moreover, small molecules can be effectively purified by a number of methods, including train sublimation. By controlling the structural molecular arrangement at the organic/dielectric interface, it is possible to tune the solid-state photophysical properties of organic thin films. For these reasons, conjugated small molecules are considered promising structures for OLETs realization. In spite of the great potential of the microscopic molecular packing control, until now the small molecules approach resulted in limited ambipolarity and light-emission efficiency.

Few examples are reported in the literature with the realization of truly ambipolar OLETs based on small molecules. However, in most of these devices, either the optical features (such as the gate-controlled emission-zone

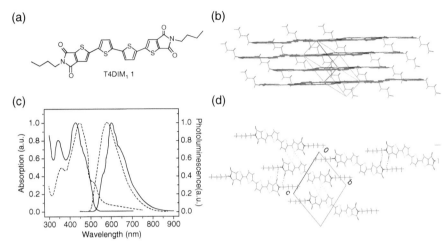

FIGURE 4.13   (a) Molecular structure of T4DIM. (b) Normalized absorption and emission of T4DIM solution (CH_2Cl_2, 4×10^{-5} M, dashed line) and film (100 nm thick, solid line). (c) Arbitrary view of the crystal packing showing the π–π stacks. (d) View along the axis of one 1D network generated by C–H⋯O interactions (dashed black). S⋯S and C–H⋯S contacts (light blue). [*Source*: Melucci *et al.* [78]. Adapted with permission of ACS Publications.]

location) are not satisfactory or the optoelectronic performances are dominated by the minority charge carriers. Adachi *et al.* [105] reported on the fabrication of blue-emitting ambipolar OLET based on the OSC 4,-4′-bis(styryl) biphenyl with charge-carrier mobility as high as 0.01 cm^2/V s. However, no images of emitting stripe and device brightness measurements were shown. In other reported examples, the optoelectronic characteristics had mainly unipolar charge-transport characteristics as in the case of ambipolar light-emitting quaterthiophene derivative (i.e., α,ω-dihexylcarbo-nylquaterthiophene) [24].

In the following, we discuss a notable example of light-emitting ambipolar small-molecule prototype for OLET realization, which was recently published [106]. The material is a new, readily available, and synthetically flexible oligothiophene derivative bearing thienopyrrolyl dione symmetric ends (2,20-(2,20-bithiophene-5,50-diyl)bis(5-butyl-5H-thieno[2,3-c]pyrrole-4,6)-dione (named T4DIM, Figure 4.13a).

The observation of the optical features of the molecule revealed that CH_2Cl_2 solution and film absorption spectra are similar, suggesting high molecular planarity and rigidity (Figure 4.13c). The absorption maximum is located at about 450 nm in solution and 430 nm in film (shoulder at 520 nm), strongly red-shifted with respect to dihexyl substituted quaterthiophene (DH4T, 402 nm) and diperfluorohexyl quaterthiophene (398 nm), but close to that of dihexyl-substituted sexithiophene (440 nm) [107], as a result of the higher conjugation length originated by the inclusion of the pyrrolidinone moieties in the conjugated system. Consistently, the photoluminescence maximum is located at 579 nm in solution and in film, strongly red-shifted with respect to that of DH4T. The PLQY value in thin film is in the range expected in the case of thiophene derivatives used for OFETs (around 5% only).

The molecule lies on a crystallographic inversion center located at the midpoint of the bond between the two thiophene units. The molecular backbone is almost planar, with the two inner thiophene rings being strictly coplanar and the dihedral angle between the thiophene and the thiophene-imide units 4.4(5)°. The crystal packing (Figure 4.13b) shows that the molecules adopt a slipped π–π stacking packing mode (interplanar distance ca. 3.51 Å, sliding 3.32 Å along the long molecular axis) instead of the herringbone structure more common for oligothiophenes [108].

The effect of end groups other than alkyl chains has also been studied. Barbarella and coworkers [109] have synthesized a series of oligothiophenes that are end-capped with dimethyl-t-butylsilyl groups. Single-crystal analysis of the ter- and quater-thiophene derivatives revealed that these materials form a dimeric face-to-face π-stacked structure in the solid state, which has been theorized to give high charge mobility in OSCs. However, the molecular packing of the corresponding thin films is likely much different than that in the bulk solid state, leading to the low mobilities.

Even in the recently synthesized dithieno-[2,3-b:3'2'-d]thiophene diimide, the molecule assumes a chair-like configuration with the two-terminal propyl chains turned out of the plane: the compound displayed a layered herringbone arrangement with two orientations in the crystals. Indeed, S··· π and C–O···π interactions were observed in each herringbone packing, with a herringbone angle of 55.7°.

In the case of T4DIM, the molecules of adjacent π–π stacks are engaged in supramolecular 1D networks running across the bc plane formed through two intermolecular C–H···O H bonds in which two oxygens in antiposition interact with one hydrogen of the carbon directly attached to the nitrogen of the imide unit of two different molecules. Moreover, intermolecular contacts involving the sulfur atoms of the fused imidothiophene rings [110] of neighbor molecules not participating in the C–H···O interactions were found. The same sulfur atoms are also engaged in weak C–H···S interactions (C–H···S, H···S 3.03 Å) with the inner hydrogens of the thiophene rings.

In Figure 4.14, the optoelectronic characteristics of OLET device implementing T4DIM as active material are reported. The fact that the electroluminescence

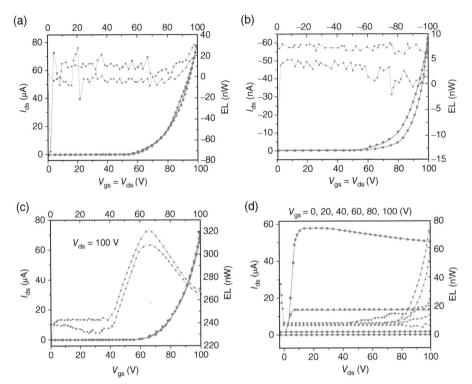

FIGURE 4.14   Electrical characteristics (blue) and corresponding light-emission intensity (purple) of single-layer OFET based on T4DIM; (a, b) n-type (a) and p-type (b) locus curves; (c) n-type transfer saturation curve; and (d) n-type multiple output curves. [*Source*: Melucci *et al.* [106]. Adapted with permission from Royal Society of Chemistry.]

peak in the optical power curve is located at about the half of the scanned gate-voltage range in the transfer characteristic is the clear indication that photons are efficiently formed and emitted in the ambipolar regime of the transistor (Figure 4.14c). Even though the mobility values are strongly unbalanced, ambipolarity in the T4DIM-based OLET is demonstrated by the n-type and p-type locus curves of Figure 4.14a and b, respectively. p-Type mobility is indeed three orders of magnitude lower than the n-type one.

It is interesting to note that even though the PLQY in thin film is not as high as thiophene-S,S-dioxide derivatives [111], but rather similar to linear end-substituted oligothiophenes usually implemented in OTFTs, the electroluminescence signal is unexpectedly high (hundreds of nanowatts as peak power) for this category of conjugated OSCs, probably due to the reduction of nonradiative decay paths for excitons.

In the T4DIM compound, the molecular core is the quaterthiophene, which is the conjugated moiety that constitutes the most-performing oligothiophene derivatives, such as the alkyl end-substituted quaterthiophenes (e.g., the dihexylquaterthiophene, DH4T) [112]. In order to promote the electron transport, the electron-withdrawing dithienothiophene-diimide moieties are fused to the quaterthiophene core. Moreover, the butyl linear chains are end-substituted for guaranteeing the necessary high-density molecular packing, suitable for efficient field-effect charge transport. As a result of the chemical strategy, a strongly n-type transport material with a residual p-type character is obtained. Despite the unbalance in the p-type and n-type character, the performance of OLETs based on T4DIM is very relevant in the scenario of single-layer single-material ambipolar OLET devices.

In order to corehend systematically the correlation between molecular structure, molecular packing, and optoelectronic performances, a class of molecules derived from the prototype T4DIM has been investigated. The modifications in the molecular structure which are related to different end-substituted units with respect to the T4DIM, mainly affect the molecular packing motif and the ambipolarity characteristics in the thin film.

The main synthetic strategy is the nonsymmetric insertion of the thieno(bis) imide (TBI) moiety into unipolar quaterthiophene derivatives. Particularly, among the plethora of possible thiophene derivatives, DH4T and perfluoro-hexylquaterthiophene (DHF4T) are used as reference model molecules for p- and n-transport in unipolar OFETs, given their successful use in multilayer OLETs [28]. One of the alkyl/perfluoroalkyl chains was replaced by a TBI group so that nonsymmetric molecules HT4N and FT4N were synthesized (Figure 4.15a and b). Moreover, quaterthiophene derivative having an unsubstituted thienyl end-ring opposite to the TBI-end was also realized for comparison, T4N (Figure 4.15c).

Interestingly, HT4N, FT4N, and T4N compounds show similar electronic density distributions of highest occupied molecular orbital (HOMO) and LUMO, while they differ from the family molecule progenitor T4DIM. Indeed, in the symmetric T4DIM, both HOMO and LUMO are delocalized along the whole molecular structure, while for nonsymmetric molecules, the LUMO is mainly localized on the TBI moiety, independently on the type of substitution opposite to the TBI.

Also, the crystal structures of the nonsymmetric compounds are similar, having in common a nearly planar backbone in a chair-like molecular arrangement (in Figure 4.15d and g, FT4N and T4N crystal structures are reported). In the crystal packing of T4N (Figure 4.15f), the molecules are aligned in an antiparallel way with respect to each other with the long molecular axis oriented along the a axis and adopt a herringbone-like arrangement with slipped

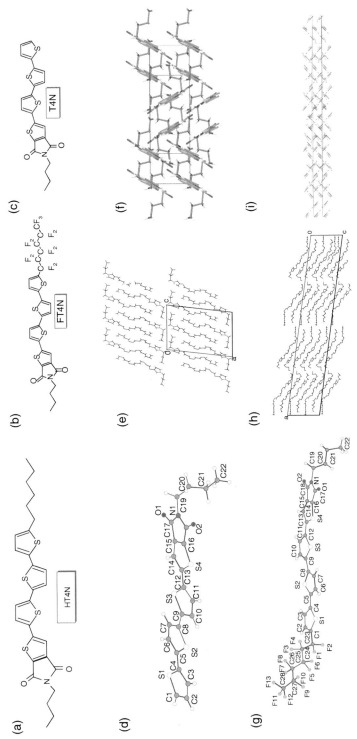

FIGURE 4.15 Molecular structure of the nonsymmetric thieno(bis)imide (TBI)-ended quaterthiophenes: (a) with alkyl chain end substitution, (b) with perfluorohexyl chain end substitution, and (c) without any end substitution. Crystal structure of T4N (d) and FT4N (g). (e, h) View down the *b*-axis of the crystal packing of T4N (top) and FT4N (down), respectively. (f, i) View along the long molecular axis of T4N (top) and FT4N (down), respectively, showing the herringbone-like packing. The H atoms, perfluorohexyl, and *n*-butyl chains in FT4N have been removed for clarity. [*Source:* Melucci *et al.* [78]. Adapted with permission of ACS Publications.]

π–π stacking. In FT4N, one of the two independent molecules shows the usual anti-anti-anti orientation of the S atoms, whereas the second one exhibits an uncommon syn-anti-anti conformation (Figure 4.15h). In the crystal packing of FT4N, the two conformers are aligned face-to-face with the long molecular axis almost parallel to the diagonal of the ac plane and establish slipped π stackings. Looking down the long molecular axis, a herringbone-like pattern is noticed (Figure 4.15i). Intermolecular C–H···O and C–H···F interactions connect the π–π stacks.

Interestingly, while compounds T4N and FT4N show herringbone-like packing, as generally observed for linear alkyl end-substituted oligothiophenes, T4DIM having two TBI end moieties, showed a slipped π–π-stacking packing motif. The "herringbone" and "π stack" structures are seen to be closely related and differ primarily in the major slip direction along both long and short molecular axes.

As a result, higher mobilities at room temperature might be achieved by designing conjugated molecules that stack face-to-face (π stack) in the solid state, thus increasing the intermolecular interaction, while maintaining the desirable 2D character of the herringbone structure [113,114]. In addition to having a 2D π-stacked morphology, the ideal OTFT material would self-assemble on the surface of a substrate, for example, the insulating layer of an OTFT gate, such that the π stacks would be parallel to the surface, that is, the direction of the highest mobility would coincide with the direction of the current flow in the OFETs.

From the standpoint of increasing charge-carrier mobility by generating large valence or conduction bandwidths (proportional to the orbital overlaps of adjacent molecules) in crystalline conjugated oligomers, it would seem that crystals with π-stacked molecules would be better than those with herringbone packing. Also, from the standpoint of the hopping theory of electron conduction, π stacks would also lead to higher mobility because the hopping rate decreases exponentially with the hop distance and increases with intermolecular orbital overlap [115].

Indeed, T4DIM shows the best p- and n-type charge mobility with respect to all nonsymmetric compounds, as expected from the molecular π stacking in thin films.

With respect to the light-emission characteristics of HT4N, FT4N, and T4DIM compounds, electroluminescence signal in the transfer characteristics is peaked where the balance between the hole and electron currents is maximized ($V_{gs} \sim 1/2V_{ds}$) (Figure 4.16a,b,c), which is a clear fingerprint of the light generation inside the OLET channel.

In the case of the T4DIM compound, the emissive stripe within the transistor channel is closer to the drain electrode as a consequence of the

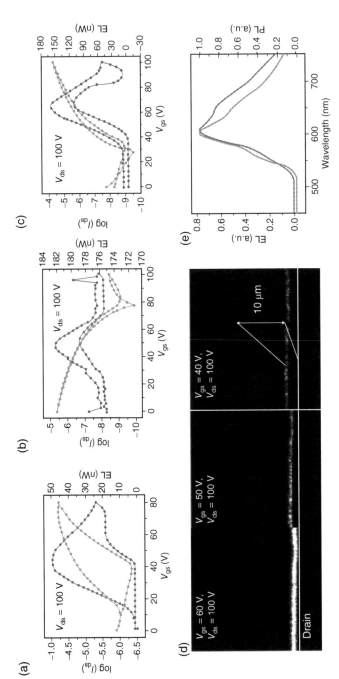

FIGURE 4.16 Optoelectronic transfer curves of an OLET device based on (a) FT4N, (b) HT4N, and (c) T4DIM molecules. The blue lines represent the measured OLET drain current, and the purple lines are the emitted photon power. (d) High-resolution images of emissive stripe shifting by decreasing the gate voltage in a working T4DIM OLET. The device channel size is 70 μm. (e) Electroluminescence spectrum of OLET based on T4DIM (red line) compared to the photoluminescence spectrum (blue line). [*Source:* Melucci *et al.* [78]. Adapted with permission of ACS Publications.]

difference between hole and electron mobilities. However, the electroluminescence generation area is well separated from the drain edge, preventing optical coupling of the emitted light with the metal electrodes (Figure 4.16d). Moreover, the EL and PL spectra are very similar indicating that the emitting state in the thin film is the same, regardless of the optical or electrical generation of excitons (Figure 4.16e).

This systematic study highlighted the possibility of one-to-one control of the multifunctional properties in thermally evaporated linear oligomers by suitable functionalization of the conjugated molecular core.

Indeed, the introduction of the TBI moiety in alkyl-substituted oligothiophenes allows to (i) tune the HOMO energy values of oligothiophenes, while maintaining the LUMO energy unchanged; (ii) localizing the LUMO distribution on the oligomer periphery; (iii) mastering the charge transport (ambipolar with predefined major p- or n-type behavior) by a proper TBI opposite end substitution; and (iv) enabling thin-film electroluminescence in combination with ambipolar charge transport.

Nonetheless, we have to observe that the most-performing optoelectronic characteristics (i.e., maximum current, current balance, and optical power) are obtained in the case of symmetric compound which displays the crystal structure motif more suitable for efficient charge transport in field-effect device.

A further fundamental study was performed to investigate the structure–property correlation in multifunctional molecular systems implemented as active material in OLETs in order to define a property-specific design strategy. As we have highlighted, TBI-based molecular semiconductors combining good processability, tunable self-assembly, ambipolar charge transport, and electroluminescence can be considered an ideal test bed for investigating the relationships between the molecular structure, packing modalities, charge mobility, and light emission in organic thin films.

Particularly, TBI materials having a fixed π-conjugated backbone (quaterthiophene, T4DIM), length, and core, but differing in the alkyl end-substituents, were recently reported in order to report a systematic investigation of structure–property correlation in thermally sublimated small-molecule thin films [115].

The experimental and theoretical results obtained in the case of linear chains (C1–C8) are highly interesting in shedding light into the role of the n-alkyl end of the TBI moiety for the selective modulation of the electroluminescence.

Indeed, contrarily to what generally observed in conventional oligothiophenes, no significant effects on the optical properties, thin-film morphology, and n-type charge mobility resulted from the different structural end motifs, while modulation in the p-type charge mobility and electroluminescence was observed.

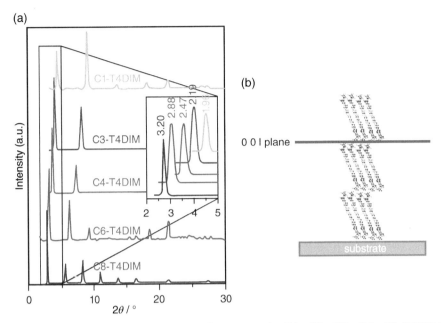

FIGURE 4.17 (a) XRD patterns of compounds C1, C3, C4, C6, C8-T4DIM devices directly exposed to the X-ray beam. In the inset, the small-angle region is enlarged and the corresponding interlayer distances are reported in nm. (b) Schematic representation of the C4-NT4N molecule arrangement on the substrate. [*Source*: Melucci *et al.* [115]. Adapted with permission of Royal Society of Chemistry.]

In thermally evaporated thin films, the molecules organize in dense layers given the strong π–π and hydrogen bond interactions. In every layer, the conjugated backbone is located in the inner part, while the alkyl chains lie on the surfaces and allow efficient packing of the molecules, then the layers pile up along the molecular long axis. The XRD profiles are very similar to one another, differing in the position of the main reflections.

Among the different molecules, the interlayer distance progressively increases with increasing length of the alkyl chain. Based on the preferential orientation parameter obtained from the derivative with C3 alkyl chain, for which the single-crystal structure is known, the sketch of the molecular organization in thin films is depicted in Figure 4.17.

Moreover, linear alkyl tail compounds show ambipolar behavior with a major n-type contribution to charge transport: indeed, the electron mobility ranges around $10^{-1}\ cm^2/V\,s$ for all the devices, while the hole mobility oscillates between 2 and 3.5 orders of magnitude lower than electron mobility.

In order to understand the correlation between the molecular structure and the variation in charge mobility for all the compounds in solid state, the

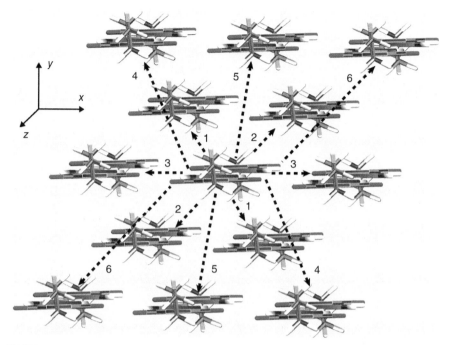

FIGURE 4.18 Charge-hopping pathway schemes for molecule C3-T4DIM in the bulk crystal. The y and z axes are perpendicular and parallel, respectively, to the molecular axis. [*Source*: Melucci *et al.* [115]. Adapted with permission of Royal Society of Chemistry.]

theoretical determination of the inner reorganization energy and the charge-transfer integral parameters within in the framework of the small polaron-hopping model was performed (see Chapter 5).

The inner reorganization energy is an intrinsic molecular property, packing independent, while the charge-transfer integral represents the electronic coupling element between neighboring molecules and therefore critically depends on the order in the solid state.

Given that the inner reorganization energy values are almost constant for the different compounds, the predominance of n-type behavior cannot be understood at a molecular structure level, but also requires the analysis of the molecular packing. Consequently, the transfer integrals of C3-T4DIM, for which XRD spectra are comparable for the single-crystal and semicrystalline thin films, were calculated to gain an insight into the packing-charge transport relationships. The possible charge-hopping pathways are shown in Figure 4.18.

The highest electronic couplings are computed for the dimers involving first next-neighboring molecule pairs, namely pathways 1, 2, and 3: only pathway 1 allows the hole transport, while all the directions are favorable for electron transport.

The calculated electron mobility shows a good agreement with experiments, while the calculated hole mobility dramatically overestimates the FET one of about three order of magnitudes. Indeed, this discrepancy may be ascribed to the difficulty in defining an effective theoretical model for properly describing different types of field-effect charge transport in real devices.

At the theory level, the charge mobility is described by Brownian motion of charge carriers with no spatial constraints, while the experimental FET mobility strictly depends on the source–drain electromagnetic field tensor, which is affected by the specific device configuration and active-layer characteristics. In this sense, while electrons can move along any crystal directions, thanks to nonnegligible transfer integrals, hole transport is significant only along one pathway. If the latter is forbidden by some specific device configuration, even a damping of orders of magnitude may become possible for the experimental hole mobility value. Thus, without considering specific paths (see Figure 4.18), a reconciliation between theoretical and experimental determination in charge mobility for both electrons and holes can be found.

Apart from this theoretical insight into the comprehension of ambipolarity in field-effect transistor based on semicrystalline small-molecule thin films, the experimental evidence is that the increase in the minority charge-carrier current (and the consequent increase in the ambipolarity balance) is directly correlated to an increase in the emitted optical power of the OLETs. Specifically, this result is achieved by implementing as active material the derivative with the longest end-substituted alkyl chain (i.e., octyl chain).

The direct one-to-one correlation between the molecular structure of moieties forming the active layer and the light-emission characteristics of the overall device is not trivial from both the theoretical and the experimental point of view. In particular, an enormous effort is required for understanding the correlation between the small-molecule or polymer structure solid-state thin-film arrangement and functional properties as a function of the thin-film process parameters. Indeed, it is mandatory to perform fundamental studies on model organic multifunctional compounds in order to cross-correlate compositional, structural, morphological, and photophysical features in organic nanostructured thin films with theoretical *ab initio* calculations. Relevant knowledge can be gained by characterizing organic thin films by means of different investigation tools such as scanning nanoprobes (AFM, scanning tunneling microscope (STM), ultrasound force microscopy (UFM)), high-resolution diffraction techniques (glancing angle X-ray diffraction (GA-XRD), low-energy electron diffraction (LEED)), X-ray, and optical time-resolved spectroscopy (X-ray photoelectron spectroscopy (XPS), optical transient absorption spectroscopy, time-resolved photoluminescence spectroscopy).

This multidisciplinary research is the path to follow in order to define a suitable synthetic strategy for achieving high-performing active materials for single-material single-layer OLETs and thus enables an application-driven new paradigm in organic optoelectronics.

4.3 CHARGE-INJECTING ELECTRODES

Charge injection at the interface between metallic electrodes and OSCs plays a crucial role in the performance of organic optoelectronic devices. OSCs differ significantly from their inorganic counterparts, primarily because they are amorphous van der Waals solids. As a result, the electronic states are highly localized, and charge transport is by site-to-site hopping. Organics can also form clean interfaces with many metals, free of interface states in the gap. Nevertheless, a significant vacuum level offset is generally found, the origins of which are not yet fully understood. Thus, the Fermi level in the organic layer and the charge-injection barriers depend directly on the interface offset. The charge-injection process is described as thermally-assisted tunneling from the delocalized states of the metal into the localized states of the semiconductor, whose energy includes contributions from the mean barrier height, the image potential, the energetic disorder, and the applied electric field. As we will discuss in detail in Chapter 5, there is no completely satisfactory analytical theory for the field and temperature dependence of the injection current, which, for well-characterized interfaces, is related to both thermionic emission and field-induced tunneling.

In this scenario, the present section is intended to provide the current understanding of the formation of the metal–organic contact and the parameters that control the injection current in organic field-effect (light-emitting) transistors. The most important (but not the only) factor that controls the charge-injection process is the energy barrier to be overcome as the charge carrier crosses the interface.

The normal starting point in the description of the interface between a metal and a semiconductor is to define the energy difference between the respective Fermi energies of the isolated materials—the so-called built-in potential. When the contact is made, equilibrium dictates that charge flows from one material to the other, until the Fermi levels align. There are two extreme cases to consider, depending on whether the transferred charge forms an interfacial dipole within the first molecular layer of the semiconductor or occupies only dopant levels in the bulk.

The latter case is described by the Mott–Schottky model [116], which states that the vacuum levels align at the interface and then the additional charge resides in a depletion zone created by ionizing donor or acceptor dopants with consequent band bending.

In many applications of OSCs, there are no deliberant dopants. As a result, the Fermi level is not easily determined, making the analysis of experimental data liable to misinterpretation. The reader is invited to refer to Chapter 5 for a detailed description of the different models used to describe charge-injection process in OSCs.

Moreover, the fabrication processes, which are used to fabricate organic field-effect devices, are far from "ideal": metal electrodes are almost poly-crystalline and rough at the relevant molecular length scale. There may be a chemical reaction between the metal and the organic material, particularly when the top electrode is deposited, and the OSC may be doped either accidentally or deliberately in order to change its transport properties. All of these phenomena will affect injection.

Even in the presence of an "ideal" interface between a clean metal surface and a van der Waals solid, the charge-injection process is complicated by the amorphous nature of the organic layer, which results in highly localized electronic states with a random distribution of energies and a low carrier mobility [117]. In an ideal OFET,[1] the source and drain contacts are ohmic, meaning that the value of the contact resistance is negligibly small in comparison with the electrical resistance of the semiconductor (i.e., the channel resistance). While this situation can be achieved in real devices, there are several practical considerations for fabricating OFETs that are not contact limited.

In crossing an OFET channel from source to drain, charge carriers are (i) injected from the source contact into the semiconductor channel, (ii) transported across the length of the channel, and (iii) extracted from the channel into the drain. These processes can be roughly thought of as three resistors in series. The resistances associated with carrier-injection and -collection steps can be grouped into the *contact resistance* (R_c), while the resistance associated with crossing the channel length in the semiconductor is termed the *channel resistance* ($R_{channel}$). Keeping the contact resistance small compared to the channel resistance is crucial for the realization of "ohmic contacts" in OFETs (i.e., for an ohmic contact, $R_c \ll R_{channel}$) [118].

If the contacts are ohmic, then they are not bottlenecks to current flow and they can provide and collect the charge carriers that can be transported by the channel under given bias conditions. Importantly, nominally high-resistance contacts can still be ohmic as long as they are able to provide and collect the current driven through an even more resistive channel. This definition also

---

[1] Throughout the book, OTFT and OFET acronyms are used to refer to the same device, reflecting the literature use of acronyms in different contexts.

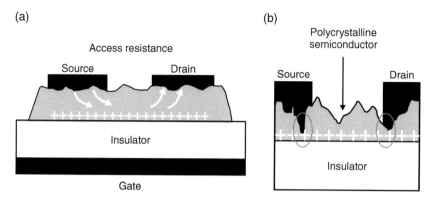

FIGURE 4.19 (a) Access resistance in a top contact OFET represented by arrows indicating the injection and extraction of charge through the bulk semiconductor film between the source/drain contacts and the conductive channel. The plus signs represent the hole accumulation layer of a p-channel device. (b) Detailed view of the contact/semiconductor interface near the accumulation layer of a top contact OFET with a very rough, polycrystalline semiconductor thin film. Ovals highlight the penetration of the top contacts deep into the film; in such devices, access resistance can often be negligible.

implies that contacts that function ohmically for a given channel length may in fact no longer be ohmic as the channel length is reduced. This is because $R_{channel}$ scales proportionally with length, so making smaller OFETs results in smaller channel resistances and reduces the upper bound on the acceptable value of contact resistance. An additional issue in OFETs is that the channel resistance is continuously lowered with increasing gate voltage, and thus, contact resistances must be low compared to the channel resistance at high gate voltage in order to be considered ohmic.

The influence of contact resistances on the performance of unipolar OFETs is extensively discussed in the literature [4,90,119], whereby the importance of the nonlinear resistances leading to diode-like current/voltage characteristics is highlighted [119]. Moreover, even the device configuration plays a key role in determining the efficiency of charge-injection process in organic field-effect devices.

Top contact OFETs generally exhibit the lowest contact resistance. This is likely because of the increased metal–semiconductor contact area in this configuration. A major contribution to contact resistance in the top contact configuration is the so-called *access resistance* (Figure 4.19a). Access resistance results from the requirement that charge carriers must travel from the source contact on top of the film down to the accumulation layer (the channel) at the semiconductor–insulator interface and then back up to the drain contact to be collected.

In order to minimize access resistance, the thickness of the OSC layer should not be too large. However, some researchers have proposed that access resistance is less than might be expected for top contact OFETs because the contact metal penetrates the film down to the accumulation layer (perhaps due to large peak-to-valley roughness of the semiconductor film or the nature of the metal deposition process) [120]. This scenario is depicted in Figure 4.19b.

In the case of the bottom contact configuration, access resistance is not an issue because the contacts are coplanar to the device accumulation channel. In addition, very small channel dimensions (W, $L < 10\,\mu m$) can be prepatterned on the insulator using conventional photolithography. However, a limitation of the bottom contact configuration is that film morphology in the vicinity of the contacts is often nonideal, that is, inhomogeneous and with defects.

It has been demonstrated that the OSC grain sizes are very small near the contacts, presumably due to heterogeneous nucleation phenomena [14]. Pentacene molecules, for example, tend to "stand up" with the long axis of the molecule perpendicular to the plane of the substrate when deposited on the commonly used insulator SiO_2 [121]. When deposited on top of gold contacts, however, strong interactions between the pentacene π-clouds and the metal surface lead to tiny grains at the contact and, in some cases, voids are observed [122].

Semiconductor growth at the complex triple interface (contact–semiconductor–insulator) is not very well understood, although it is clear that the bottom contact configuration almost always creates greater contact resistance than in the case of top contacts.

Of the two top gate OFET architectures, the top contact/top gate configuration is the more favorable one because bottom contact/top gate devices suffer from access resistance. However, it should be noted that both top gate architectures suffer from additional issues related to semiconductor top surface roughness (since this is where the channel will form) with the consequent impossibility to create a sharp interface between the insulator and the upper OSC film surface. Regarding the alignment of the top gate contact to the OFET channel in top gate devices, care must also be taken to ensure that the gate overlaps completely across the entire length (L) of the device. If the length of the gate electrode is less than the channel length or if the gate is simply misaligned, additional contact resistance will be introduced as a result of ungated semiconductor regions at one or both of the contacts [123].

It is interesting to note that the most-performing OLETs reported in the literature until now in terms of efficiency and brightness are fabricated in device configurations where access resistance is relevant. Indeed, the best-performing devices are bottom gate/top contact in the case of active regions made of multilayer small molecules [28] and are bottom contact/top gate in the case of active regions made of single-layer polymers [27].

In order to achieve an ambipolar OFET behavior, the injection and transport of both charge carrier types are required. As we have reported in detail in Chapter 3, in the first ambipolar OLET, the balance of the electron and hole currents was achieved by stacking the n- and p-transport semiconductors in a vertical heterojunction and using the same metal for the injection of both charge-carrier types [77].

Mandatory for allowing small electron-injection barrier was the use of calcium as metal for the electrodes. Schmechel *et al.* [124] showed that the use of calcium for the electron-injecting contact and gold for the hole-injecting contact in one device, pentacene as semiconductor, and an appropriate dielectric lead to an ambipolar charge transport and comparable mobilities for electrons and holes in the transistor channel.

Although in such an ambipolar OFET, electrons and holes recombine in the channel, no light emission was observed, given that efficient luminescence quenching process is prevailing in pentacene [10]. Later, Schidleja *et al.* [125] demonstrated a weak light emission arising from the charge-carrier recombination zone in pentacene and showed that, in agreement with the ambipolar current–voltage characteristics, the position of the recombination zone in the transistor channel can be tuned by the applied voltages.

As we have underlined in the previous paragraph, the use of OSCs with adequate luminescence properties is essential for achieving brighter electroluminescence from the transistor channel [69]. By using highly emissive polymeric semiconductors as active layer, efficient ambipolar injection is achieved by implementing the "two-color" electrode geometry [81], where the channel region of the transistor is defined by a low-work-function metal on one side and a high-work-function metal on the opposite side. In the case of ambipolar transistor, the nomenclature of source and drain contacts commonly used for unipolar transistors is ambiguous. Indeed, when the device is biased in a way that both electrons and holes are flowing in the channel, charges (of opposite sign) are injected from both high- and low-work-function contacts. If we define the source contact to be the one held at ground potential, the high- (or low-)work-function contact should be referred as source (drain) electrode for negative gate bias and vice versa for positive gate bias.

Light emission in the channel region of an OFET using the two-color electrode geometry was demonstrated in the first stage of the OLET technological development [126–128].

To fabricate the devices, a heavily doped n-type silicon wafer was used as the gate electrode. The gate electrode was coated with 400 nm of silicon nitride (SiNx) deposited by plasma-enhanced chemical vapor deposition.

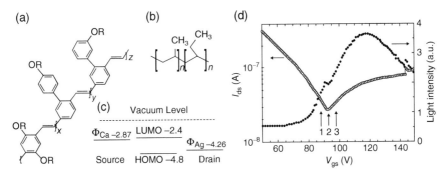

FIGURE 4.20 Molecular structure of (a) SY and (b) PPcB. (c) Energy level diagram for the Ca source electrode/SY/Ag drain electrode device structure. (d) Transfer scan characteristic for the OLET based on SY along with the corresponding emitted light intensity versus the applied gate voltage. The emission zone is located in the channel: (1) near the Ca source electrode, (2) near the center of the channel, and (3) near the Ag drain electrode. [*Source*: Swensen *et al.* [81]. Adapted with permission of AIP Publishing LLC.]

The SiNx was passivated at 190 °C with a thin film of polypropylene-*co*-1-butene, 14 wt% 1-butene (PPcB; Figure 4.20). The "Super Yellow" (SY), a polyphenylene vinylene derivative, is used as active layer.

The realization of a top-contact two-color electrode geometry was possible by developing a tilted-angled evaporation technique using a silicon shadow mask. The shadow mask was fabricated by etching two parallel rectangles ($1000 \times 100 \, \mu m^2$) through a silicon wafer (250 mm thick) separated by a 20 μm "stripe." The rectangles defined the electrode area, while the "stripe" defined the channel region. As shown in Figure 4.21, by evaporating at an angle, the shadow created by the stripe enabled the fabrication of the top-contact two-color electrode geometry. This inexpensive and clever sublimation scheme allowed the fabrication of devices with channel length of less than 5 μm.

In Figure 4.20, the transfer characteristics, together with the gate-dependent electroluminescence profile, are reported. The transfer scan was run with a constant drain voltage of 200 V, while varying the gate voltage from 0 to 200 V.

The light intensity begins increasing at around 80 V, while the hole current still dominates, reaching a maximum around 120 V, well into the electron-dominated current regime. By employing the two-color electrode geometry, light emission should be observed when both electron and hole currents are simultaneously present during device operation as displayed by the electroluminescence profile.

However, the light intensity peaks at gate voltage of 120 V, that is, when the electron current is greater than the hole current. A higher electron current

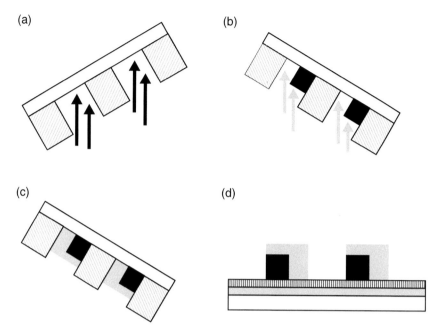

FIGURE 4.21 Schematic of the fabrication of the top-contact two-color electrode geometry by the angled evaporation technique: (a) The substrate is mounted on a silicon shadow mask at an angle to the metal sources, and the first metal, Ca, is evaporated; (b) the angle of the mask is changed with an electric motor, and the second metal, Ag, is evaporated; (c) the deposition of the two-color electrodes is complete; and (d) the final structure of the two-color OLET is obtained after the removal of the silicon shadow mask. [*Source*: Swensen *et al.* [81]. Adapted with permission of AIP Publishing LLC.]

might be necessary in ambipolar conditions in order to achieve maximum brightness given that the higher density of electron traps would reduce the number of electrons available for recombination.

Indeed, at that early stage of development of OLET technology, the incontrovertible experimental evidences of the distinguishing OLET optoelectronic features were achieved by using OSCs (in particular, conjugated polymers) presenting both charge transport properties in field-effect configuration and the preservation of efficient photoluminescence. Nonetheless, the fabrication paradigm of using (i) different metal electrodes for injecting efficiently holes and electrons in top contact configuration and (ii) surface-functionalized dielectrics for avoiding charge trapping at the dielectric–semiconductor interface was adopted by different research groups.

In the reference work published by the Sirringhaus group [22], the spatially resolved light emission in OLETs was demonstrated by using a cross-linked

BCB derivative as a buffer dielectric to prevent the trapping of electrons and calcium and gold electrodes for injecting electrons and holes. The metal deposition configuration is similar to the one showed by Swensen *et al.* [81] (two-color electrodes). However, the active material used in this case is a solution-processed conjugated polymer, poly(2-methoxy-5-(3,7-dimethyl-octoxy)-*p*-phenylene-vinylene) (OC_1C_{10}-PPV or MDMO-PPV), which has higher field-effect mobility values for electrons with respect to SY (i.e., 3×10^{-3} cm^2/V s vs 6×10^{-5} cm^2/V s).

However, also in this case, the electron and hole mobility values are still unbalanced by about one order of magnitude, and this unbalance is reflected in the potential profile/electric field along the channel and in the position of the recombination zone. Indeed, by imaging the spatially localized emission zone, it was observed that at low $|V_{gs}|$ and high $|V_{ds}|$, the light emission is largely located in the direct vicinity of the hole-injecting gold electrode. This allowed two possible explanations for the decreased quantum efficiency in this regime. A significant fraction of the formed excitons are likely quenched by the metal of the electrode or by metal atoms that diffuse into the channel close to the electrodes, as it is typically expected in the top contact configuration. Another effect relates to the increased number of electrons that could escape into the source electrode without recombining, if the distance between the recombination zone and the electrode is sufficiently small. These electrons will contribute to the drain current but not to the emitted light.

As we have discussed in detail in the previous paragraph, the effective improvement in both the EQE and the brightness in single-layer polymeric OLET was achieved by changing both the active material (F8BT instead of OC_1C_{10}-PPV) and the device architecture (bottom contact/top gate instead top contact/bottom gate).

The bottom contact/top gate configuration that was adopted presented numerous advantages with respect the bottom gate one. It does not rely on shadow mask evaporation of two dissimilar, low- and high-work-function metals, but allows the definition of electron- and hole-injecting contacts from a single metal patterned by photolithography. This improves the environmental shelf life of the device, because no reactive, low-work-function metals are required to achieve electron injection. Furthermore, the top gate structure allows more flexible use of different polymer gate dielectrics that do not need to be cross-linked and are simply solution-coated onto the semiconducting layer from orthogonal solvents.

However, the use of the same metal for the injection of electrons and holes leads to inevitable misalignment between the energy levels for at least one of the carriers resulting in large injection barriers and reduced performance in ambipolar devices.

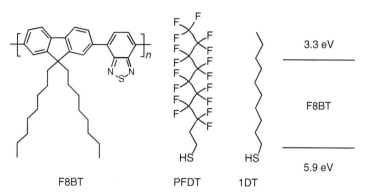

FIGURE 4.22 Chemical structure of F8BT, PFDT, and 1DT and energy level diagram of F8BT. [*Source*: Cheng *et al*. [84]. Adapted with permission from Wiley.]

In order to achieve better injection properties, the electronic properties of the metal electrodes have to match the properties of the OSC. Yet, due to the high ionization potential of F8BT of 5.9 eV [42], no appropriate metal is available to form a potentially small barrier for hole injection.

Several interface modification methods have been reported to improve charge injection. Interfacial doping of the OSC leads to big improvement of device performance in OLEDs [129] but needs to be performed selectively in the vicinity of the contacts in an OLET in order not to degrade the off-state current of the device [130,131].

Another approach is to modify the metal work function Φ by inserting dipolar molecules with well-aligned dipole moments at the interface. This can be achieved, for example, by depositing a thiol-based SAM on the metal surface prior to deposition of the OSC [132,133]. Alkanethiols and perfluorinated alkanethiols are well-known molecules that form uniform and ordered SAMs on gold or silver surface to control metal Φ. Since they have opposite electric dipole moments associated with the molecule and the Au–S bond at the interface, they modify the gold work function in different directions [134].

Thus, a possible route for reducing contact resistance in F8BT-based OLET using gold electrodes for injecting both electrons and hole is to functionalize source and drain bottom contact with different SAMs.

Particularly, Cheng *et al*. [84] used two different SAMs for increasing and decreasing gold work function, respectively, 1H,1H,2H,2H-perfluorodecanethiol (PFDT) and 1-decanethiol (1DT) (Figure 4.22).

Understanding the effect of SAM contact treatments on ambipolar charge injection requires taking into account a range of different factors and cannot be explained simply by assuming SAM-induced changes of the metal work function.

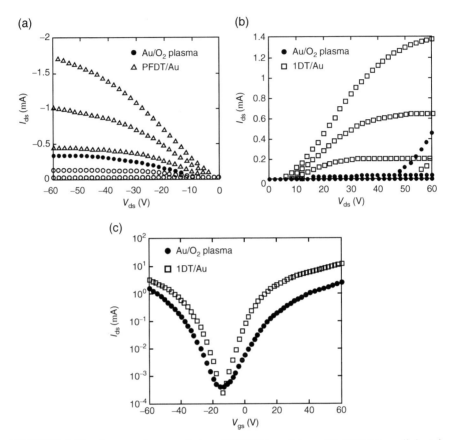

FIGURE 4.23 Output characteristics of F8BT transistor with PMMA as dielectric ($C_i = 4.0\,nF/cm^2$, $L = 10\,\mu m$, $W/L = 1000$) for different V_{gs} (from 0 to –60V and from 0 to 60V in steps of ±10V); (a) p-type curve obtained from O_2 plasma-treated Au (closed circles) and PFDT/Au contact (open triangles); (b) n-type curve of O_2 plasma-treated Au (closed circles) and 1DT/Au contact (open squares); (c) transfer characteristics of O_2 plasma-treated Au (closed circles) and 1DT/Au (open squares) at $V_{ds} = -60\,V$. [*Source*: Cheng *et al.* [84]. Adapted with permission from Wiley.]

According to the authors, the main factor, which was found to be responsible for the good ambipolar charge-injection properties of 1DT-treated gold electrodes, was a reduced thickness of the spin-coated polymer semiconductor film on top of the gold electrodes.

Compared to O_2 plasma-treated gold, the overall current is increased in both PFDT/Au and 1DT/Au. The improvement in the p-type characteristics in the case of PFDT/Au and the n-type characteristics in the case of 1DT/Au is reported in Figure 4.23.

The hole current of PFDT/Au device is larger by a factor of 5 and the electron current of 1DT/Au device by one order of magnitude with respect to that of O_2 plasma-treated gold devices. The suppression of the current at small source–drain voltages is also reduced, reflecting a smaller contact resistance.

The extracted saturation mobilities are increased to values of 1×10^{-3} cm²/V s for holes (PFDT/Au) and electrons (1DT/Au), respectively. The threshold voltages for hole accumulation (PFDT/Au) and electron accumulation (1DT/Au) are also significantly reduced. These observations are all fully consistent with the expected SAM-induced increase of work function and associated lowering of the hole-injection barrier in the case of PFDT/Au and the corresponding reduction of work function and lowering of the electron-injection barrier in the case of 1DT/Au.

What is more surprising is that in the case of 1DT/Au, the improvements in the device characteristics for operation was observed not only in the electron accumulation mode but also in the hole accumulation mode, when compared with O_2 plasma-treated gold contacts.

The ambipolar transfer characteristics shown in Figure 4.23 also show very clearly the simultaneous enhancement of both the electron and hole currents by 1DT treatment of the electrodes compared to O_2 plasma-treated gold. This behavior is surprising because the lowering of the work function by 1DT is expected to increase the hole-injection barrier significantly (by about 0.5 eV if one assumes that the Mott–Schottky limit holds). In contrast, the electron current of the PFDT/Au devices is strongly suppressed, as expected from the high work function of the PFDT/Au electrodes.

To obtain direct information about the charge-injection barriers at the metal–semiconductor interface, which are known to often deviate from the simple Schottky–Mott limit [120], spectroscopic techniques are typically used. Particularly, ultraviolet photoemission spectroscopy (UPS) is used to determine the energy offset between the Fermi level of the injecting metal and the valence band edge (the band derived from HOMO level) of the OSC according to the different interfacial functionalization. In Figure 4.24, the UPS spectra in the valence band region of F8BT films on top of Au with UV/ozone (equivalent to O_2 plasma treatment), PFDT/Au, and 1DT/Au are reported.

From the collected spectra, it is possible to determine the energy diagram by locating the gold Fermi level (E_F) and the edge of the first delocalized valence band of F8BT (VB$_{onset}$). The difference between E_F and VB$_{onset}$ should reflect the interface injection barrier encountered by holes injected from gold into F8BT.

In the case of devices with clean gold (UV/ozone and O_2 plasma-treated gold surfaces are expected to be chemically equivalent), the hole-injection barrier (HIB) is about 0.6 eV. For PFDT/Au, the HIB is very similar, that is, about 0.7 eV. This is most likely caused by pinning of the Fermi level at the interface at the polaron level of F8BT for substrate work-function values greater than 5.2 eV [135,136]. However, it is clear that for the 1DT/Au device,

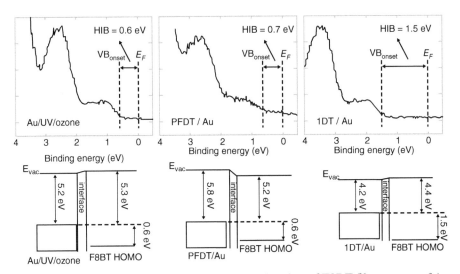

FIGURE 4.24 UPS spectra in the valence band region of F8BT films on top of Au with UV/ozone (equivalent to O_2 plasma treatment), PFDT/Au, and 1DT/Au. The Fermi level (E_F) position of Au is indicated in the figures. The hole-injection barrier (HIB) is determined between the Fermi level (E_F) and the first delocalized valence band edge (VB$_{onset}$). Intensity is given in arbitrary units. The energy diagrams regarding the injection barrier of holes at the interface are also shown. [*Source:* Cheng *et al.* [84]. Adapted with permission from Wiley.]

the HIB is dramatically increased to 1.5 eV, which is almost twice that of UV/ozone-treated gold and PFDT/Au.

These results demonstrate that the charge-injection barriers of the interfaces do not follow the simple Schottky–Mott limit; instead, significant interface dipoles were found, and pinning of the substrate Fermi level was observed for UV/ozone-treated gold and PFDT/Au.

However, these results cannot explain the finding that although the hole-injection barrier at the 1DT/Au is increased significantly, the hole contact resistance is reduced.

As mentioned earlier, one possible factor deemed responsible for the good ambipolar charge-injection properties of 1DT-treated gold electrodes is the reduced thickness of the spin-coated polymer semiconductor film on top of the gold electrodes.

Indeed, it is well-known that the contact resistance of organic transistors in the staggered configuration is expected to be very sensitive to the thickness of the OSC since charges after being injected at the interface need to be transported through the bulk of the film, before reaching the accumulation layer at the interface.

It was found that the 1DT/Au treatment leads to a reduction of film thickness on top of the electrodes by a factor of 2 compared to O_2 plasma-treated gold, probably due to the associated

SAM-induced changes of the electrode surface energy. This leads to a lowering of the contact resistance associated with transport through the bulk of the F8BT film, which more than compensates for the increase of injection barrier by about 0.5 eV.

Given that controlling the energetics, chemical doping, structure, and morphology at the metal/OSC interface is a complex and intercorrelated task, it is not possible to exclude that SAM-induced changes in the molecular packing and orientation of the polymer semiconductor might also contribute to the good ambipolar charge injection from 1DT/Au contacts into the polymer semiconductor.

From a general point of view, SAMs such as thiols are not compatible with devices that require elevated postdeposition annealing temperatures, which are needed to achieve good and reliable performances in numerous conjugated polymers, including many polyfluorenes (such as F8BT) [92,137,138]. The bond is not strong enough, and the SAM is detached from the gold surface under these conditions. Moreover, solution deposition of the SAMs makes it difficult to achieve asymmetric electrodes.

Moreover, in addition to high-temperature stability, the materials should be compatible with high-resolution patterning techniques such as photolithography. When photolithography is used with air-stable electrode materials such as gold, ambipolar OFETs commonly suffer from high contact resistance and, hence, large threshold voltages for electrons. Finally, the functionalization materials are required to possess suitable energy levels for electron injection, if air-stable electrode such as gold is used for hole injection.

Air-stable transparent metal oxides such as zinc oxide (ZnO) with a reported electron affinity of 3.7 eV [139] fulfill all these requirements and have recently been employed in vertical diode structures [139–141] using solution-processing techniques such as spin-coating [142] and spray pyrolysis [139,140,143].

Gwinner et al. [144] used ZnO as an n-type metal oxide to facilitate electron injection into F8BT-based OLETs in the usual top gate/bottom contact configuration.

Particularly, they deposit a 20–30 nm thin ZnO layer between the gold source–drain electrodes ($L = 20 \mu m$, $W = 4 mm$) and the semiconductor by using different wet techniques.

ZnO films were either deposited by spray pyrolysis at 350 °C in air using zinc acetate dihydrate in methanol or by spin-coating in nitrogen atmosphere from precursor solutions (0.1–0.3 M).

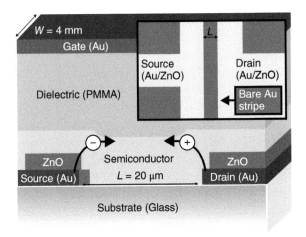

FIGURE 4.25 Illustration of a top gate/bottom contact ambipolar FET with patterned ZnO as electron injector. The inset shows a microscopy image of the substrate with gold and ZnO electrodes. [*Source*: Gwinner *et al.* [144]. Adapted with permission from Wiley.]

To pattern the metal oxide, positive photoresist was redeposited and photolithography was performed using the reversal copy of the initial electrode photomask pattern translated by typically 2–3 μm. This intentional misalignment leads to the electron source electrode being completely covered with the metal oxide, whereas the hole source electrode exhibits a bare gold stripe adjacent to the channel, which can be used as the hole-injecting electrode. After the resist development, the unprotected metal oxide areas inside the channel were then etched in HCl diluted with distilled water (ZnO: 1:900). In Figure 4.25, a schematic of the device is reported together with a microscopy image of the substrate where bare gold and ZnO electrodes are visible.

Given the asymmetry of the OLET structure, the hole transport remains essentially unchanged with a mobility of $\mu_h = 1.4 \times 10^{-3}$ cm^2/V s and a threshold voltage for holes of $V_{t,h} = -30$ V with respect to the nonfunctionalized reference device. However, the electron current is significantly enhanced. The electron mobility remains constant ($\mu_e = 6 \times 10^{-4}$ cm^2/V s), but the threshold voltage for electrons is dramatically lowered from $V_{t,e} = 34$ V for bare gold to only 5 V with ZnO.

As a consequence, the charge transport becomes more balanced, and in the ambipolar regime, in which light emission occurs, the drain current is about 20 times higher (Figure 4.26a).

The current in the linear operation regime is significantly less suppressed for the ZnO-modified contacts than for bare gold electrodes, which is considered

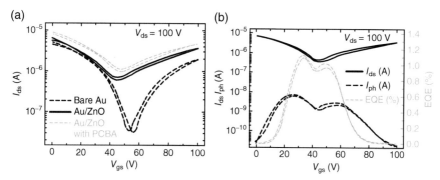

FIGURE 4.26   (a) Transfer characteristics of F8BT OLETs ($L = 20\,\mu m$, $W = 4\,mm$, 450 nm PMMA, 10 nm gold gate) with bare gold electrodes and ZnO-modified electron source (without and with PCBA). (b) Drain current I_{ds}, photocurrent I_{ph} induced in a photodiode on top of the OLET, and EQE of the F8BT OLET with Au/ZnO electrodes during a transfer scan. [*Source*: Adamopoulos *et al.* [143]. Adapted with permission of AIP Publishing LLC.]

as evidence of the possible reduction of the contact resistance. However, the electrical characteristics are still not ideal, which is at least partly due to the contribution from the high bulk resistance in the F8BT layer.

It is possible to further increase the performance by modifying the ZnO with a monolayer of phenyl-C61-butyric acid (PCBA) [145,146]. In addition to a slightly higher electron mobility ($\mu_e = 7 \times 10^{-4}\,cm^2/V\,s$), the threshold voltage is further lowered to $V_{t,e} = 1\,V$. The enhanced hydrophobicity induced by the ZnO surface functionalization of the fullerene derivative with a contact angle of $64°$ compared to $19°$ for bare ZnO is the cause of this further improvement.

As we discussed previously, the increase in hydrophobicity results in a reduction of the thickness of the subsequently deposited F8BT, leading to lower bulk resistance [90] and threshold voltages [92] for both charge carrier types.

The considerable improvement of the ambipolar current by more than one order of magnitude is crucial for increasing the singlet exciton density in the recombination zone of the OLET, which is directly correlated to the brightness. The balanced ambipolar charge transport with ZnO as electron-injection layer is further demonstrated in Figure 4.26b, where the simultaneously recorded photocurrent I_{ph} of the emitted photons on top of the device is plotted. The extracted EQE of the OLET is almost constant throughout the very broad ambipolar regime at a value of about 1% [104], which is underestimated given the fact that only the part of the emitted light is collected that exits the device through the semitransparent metal electrode on the top.

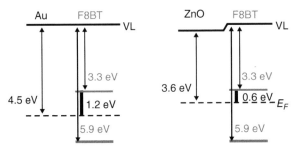

FIGURE 4.27 Derived energy level diagrams at the gold/F8BT and ZnO/F8BT interfaces (VL: Valence Level). [*Source*: Gwinner *et al.* [144]. Adapted with permission from Wiley.]

The reduction of the injection barrier by a factor of 2 is considered as the reason for the improved electron currents in the F8BT OLETs with ZnO-modified electrodes.

UPS was used for gaining information about the energy level alignment at the F8BT/ZnO interface.

In the case of a bare gold substrate, the secondary low-energy electron cutoff remains unchanged upon F8BT deposition, implying vacuum level alignment and a work function of 4.5 eV for both the substrate and the polymer film. The situation is different when the F8BT is deposited on top of ZnO. Due to the relatively low work function of ZnO, measured to be 3.6 eV, an interfacial dipole is observed as a consequence of electron transfer to the F8BT, which increases the effective work function to 3.9 eV.

Combining the UPS observations and using a 2.6 eV band gap of F8BT [42,118], the electron-injection barrier into the F8BT LUMO can be estimated to be 1.2 eV in the case of a bare gold substrate, and only 0.6 eV in the case of a ZnO-modified electrode. The energy level diagrams for both substrates are summarized in Figure 4.27. It should be emphasized that the barrier reduction by 0.6 eV down to the minimum of 0.6 eV includes the effect of the interface dipole of 0.3 eV and would be even higher without it.

Quantitatively, the observed electron current increase, which is due to a dramatically reduced threshold voltage, cannot be predicted simply from a thermionic emission model with a given injection barrier [118]. This might reflect the presence of disorder, which weakens the exponential dependence of the injection current on barrier height, owing to the availability of transport states in the tail of the density of states [147]. Moreover, the more complex device characteristics including current crowding [91] to assist the injection, as well as significant bulk resistance [90], may contribute.

As we have observed in the case of electrode functionalization by SAMs, the electron/hole currents in FETs in staggered device configuration are dominated by the injection into the active material and the transport through the channel, as long as the bulk resistance is not significantly enhanced by the functionalization layer.

This hypothesis is excluded in the case of ZnO given that the use of different deposition techniques for depositing ZnO layer modulates the maximum electron current achievable. Particularly, the OFETs with sprayed ZnO show the highest electron currents, but the corresponding ZnO films show lower conductivity than the high-temperature spin-coated ZnO films.

It is evident that the main factor that distinguishes ZnO films prepared under different conditions in terms of their performance as electron-injection layers is their surface roughness. Therefore, high-quality films with minimized surface roughness are desirable for best performance. This technique was claimed to be technologically relevant and advantageous given its use for optimizing n-type current performances in blue-emitting ambipolar OLETs, which presented air-stable electrodes.

Using air-stable electrodes, such as gold, typically can only ensure decent electron injection in materials with rather low band gap, and so far, it has not been possible to balance the hole and electron transport in materials with band gap toward the blue end of the visible spectrum without using unstable metals such as calcium or aluminum.

A polyfluorene derivative, namely F8, poly(9,9-dioctylfluorene), with a very low energy LUMO level (~2.6 eV) was used as active material in devices with ZnO electrode [144]. Electron currents are several orders of magnitude lower than the hole currents. Indeed, F8-based OLETs showed blue light emission pinned in the vicinity of the electron-injecting electrode given the unfavorable electron injection from gold into the device channel.

Applying ZnO as electron-injection layer, however, leads to an increase of the electron currents by typically two orders of magnitude and therefore to nicely balanced charge transport.

The extracted electron mobility is 1×10^{-3} cm^2/V s, and $V_{t,e}$ is substantially reduced from 68 to only 8 V. The current in the ambipolar regime and, thus, the light intensity also dramatically increase by two orders of magnitude.

As we have described throughout this section, the comprehension of the complex processes taking place at the contact between metal and OSC layers when a voltage bias is applied is not trivial. Indeed, no straightforward correlation between charge-injection and light formation processes can be derived even though light emission intensity and location within the channel in OLET model devices were used as probe for characterizing the contact formation between a semiconductor and a metal contact.

Moreover, the measurement of the energy barriers for charge injection is complicated by the failure of the Mott–Schottky rule of vacuum level alignment: some of the built-in potential is accounted for by an interfacial dipole. An additional complication arises because the energy levels in the organic layer are not always well defined and may change due to the introduction of charged impurities as the organic layer is deposited. The HOMO level is easily determined by UPS experiments, but its energy must be adjusted according to slow polarization effects before it can be used to compute the hole barrier. The LUMO energy level is even less easily determined, especially when it is derived from the HOMO energy level using the optical gap, which has to be measured from optical spectra.

4.4   CONCLUSIONS

In this chapter, the attention was focused on the principal constituting elements of the OLET device, namely the dielectric layer, the emissive channel, and the injecting contacts. It is worth highlighting that, although the structural and functional features of each element were analyzed independently, their behavior is strictly interrelated and they collectively determine the device characteristics. Indeed, the OLET should be considered as a device platform, whose specific architectural features impose requirements and constrains on the individual device subcomponents. The charge transport and photophysical processes taking place in OLETs are described and discussed in Chapter 5.

REFERENCES

[1] L. L. Chua, J. Zaumseil, J. F. Chang, E. C. W. Ou, P. K. H. Ho, H. Sirringhaus, and R. H. Friend, *Nature* **2005**, *434*, 194.

[2] W. W. A. Koopman, S. Toffanin, M. Natali, S. Troisi, R. Capelli, V. Biondo, A. Stefani, and M. Muccini, *Nano Lett.* **2014**, *14*, 1695.

[3] F. Dinelli, M. Murgia, P. Levy, M. Cavallini, and F. Biscarini, *Phys. Rev. Lett.* **2004**, *92*, 116802.

[4] H. Sirringhaus, *Adv. Mater.* **2005**, *17*, 2411.

[5] A. Facchetti, M. H. Yoon, and T. J. Marks, *Adv. Mater.* **2005**, *17*, 1705.

[6] G. Wilk, R. Wallace, and J. Anthony, *J. Appl. Phys.* **2001**, *89*, 5243.

[7] G. Horowitz, *Adv. Mater.* **1998**, *10*, 365.

[8] Peng, X. *et al.*, *Appl. Phys. Lett.* **2013**, *57*, 1990.

[9] G. Horowitz, *J. Mater. Res.* **2004**, *19*, 1946.

[10] A. Volkel, R. Street, and D. Knipp, *Phys. Rev. B.* **2002**, *66*, 195336.

[11] S. Kasap, *Principles of Electrical Engineering Materials and Devices* (Irwin–McGraw–Hill, Boston, **1997**).

[12] R. Wallace and G. Wilk, *MRS Bull.* **2002**, 192.

[13] M. Ohring, *Engineering Materials Science* (Academic Press, San Diego, **1995**).

[14] C. Boalas, *et al.*, *Phys. Rev. Lett.* **1996**, *76*, 4797.

[15] J. Collet and D. Vuillaume, *Appl. Phys. Lett.* **1998**, *73*, 2681.

[16] M. H. Yoon, *et al.*, *J. Am. Chem. Soc.* **2005**, *127*, 10388.

[17] S. Steudel, *et al.*, *Appl. Phys. Lett.* **2004**, *85*, 4400.

[18] C. Kim, A. Facchetti, and T. J. Marks, *Science* **2007**, *318*, 76.

[19] C. J. Drury, C. M. J. Mutsaers, C. M. Hart, M. Matters, and D. M. De Leeuw, *Appl. Phys. Lett.* **1998**, *73*, 108.

[20] H. Sirringhaus, T. Kawase, R. H. Friend, T. Shimoda, M. Inbasekaran, W. Wu, and E. P. Woo, *Science* **2000**, *290*, 2123.

[21] L. L. Chua, P. K. H. Ho, H. Sirringhaus, and R. H. Friend, *Adv. Mater.* **2004**, *16*, 1609.

[22] J. Zaumseil, R. H. Friend, and H. Sirringhaus, *Nat. Materials*, **2006**, *5*, 69.

[23] F. Todescato, R. Capelli, F. Dinelli, M. Murgia, N. Camaioni, M. Yang, R. Bozio, and M. Muccini, *J. Phys. Chem. B* **2008**, *112*, 10130.

[24] R. Capelli, F. Dinelli, S. Toffanin, F. Todescato, M. Murgia, M. Muccini, A. Facchetti, and T. J. Marks, *J. Phys. Chem. C* **2008**, *112*, 12993.

[25] P. Mei, M. Murgia, C. Taliani, E. Lunedei, and M. Muccini, *J. Appl. Phys.* **2000**, *88*, 5158.

[26] M. Muccini, M. Murgia, F. Biscarini, and C. Taliani, *Adv. Mater.* **2001**, *13*, 355.

[27] M. C. Gwinner, D. Kabra, M. Roberts, T. J. K. Brenner, B. H. Wallikewitz, C. R. McNeill, R. H. Friend, and H. Sirringhaus *Adv. Mater.* **2012**, *24*, 2728.

[28] R. Capelli, S. Toffanin, G. Generali, H. Usta, A. Facchetti, and M. Muccini, *Nat. Materials* **2010**, *9*, 496.

[29] L. A. Majewski, M. Grell, S. D. Ogier, and J. Veres, *Org. Elect.* **2003**, *4*, 27.

[30] L. A. Majewski, R. Schroeder, and M. Grell, *Adv. Mater.* **2005**, *17*, 192.

[31] F. C. Chen, C. W. Chu, J. He, Y. Yang, and J. L. Lin, *Appl. Phys. Lett.* **2004**, *85*, 3295.

[32] H. G. O. Sandberg, T. G. Backlund, R. Osterbacka, and H. Stubb, *Adv. Mater.* **2004**, *16*, 1112.

[33] S. Takebayashi, S. Abe, K. Saiki, and K. Ueno, *Appl. Phys. Lett.* **2009**, *94*, 083305.

[34] S. P. Tiwari, J. Kim, K. A. Knauer, D. K. Hwang, L. E. Polander, S. Barlow, S. R. Marder, and B. Kippelen, *Org. Electron.* **2012**, *13*, 1166.

[35] D. K. Kim, Y. M. Lai, T. R. Vemulkar, and C. R. Kagan, *ACS Nano* **2011**, *5*, 10074.

[36] J. Veres, S. D. Ogier, S. W. Leeming, D. C. Cupertino, and S. M. Khaffaf, *Adv. Func. Mater.* **2003**, *13*, 199.

[37] J. Veres, S. Ogier, G. Lloyd, and D. de Leeuw, *Chem. Mater.* **2004**, *16*, 4543.

[38] R. C. G. Naber, M. Bird, and H. Sirringhaus, *Appl. Phys. Lett.* **2008**, *93*, 023301.

[39] D. A. Hucul and S. F. Hahn, *Adv. Mater.* **2000**, *12*, 1855.

[40] T. Furukawa, *Phase Transitions* **1989**, *18*, 143.

[41] T. P. Ma and J. P. Han, *Electron Device Lett.* **2002**, *23*, 386.

[42] J. Zaumseil, C. L. Donley, J. S. Kim, R. H. Friend, and H. Sirringhaus, *Adv. Mater.* **2006**, *18*, 2708.

[43] N. H. Hansen, C. Wunderlich, A. K. Topczak, E. Rohwer, H. Schwoerer, and J. Pflaum, *Phys. Rev. B* **2013**, *87*, 241202.

[44] K. J. Baeg, D. Khim, S. W. Jung, M. Kang, I. K. You, D. Y. Kim, A. Facchetti, and Y. Y. Noh, *Adv. Mater.* **2012**, *24*, 5433.

[45] L. Jinhua, L. Danqing, M. Qian, and Y. Feng, *J. Mater. Chem.* **2012**, *22*, 15998.

[46] B. N. Pal, B. M. Dhar, K. C. See, and H. E. Katz, *Nat. Mater.* **2009**, *8*, 898.

[47] Y. Ha, J. D. Emery, M. J. Bedzyk, H. Usta, A. Facchetti, and T. J. Marks, *J. Am. Chem. Soc.* **2011**, *133*, 10239.

[48] Y. Ha, S. Jeong, J. Wu, M.-G. Kim, V. P. Dravid, A. Facchetti, and T. J. Marks, *J. Am. Chem. Soc.* **2010**, *132*, 17426.

[49] J. T. Ye, S. Inoue, K. Kobayashi, Y. Kasahara, H. T. Yuan, H. Shimotani, and Y. Iwasa, *Nat. Mater.* **2010**, *9*, 125.

[50] S. Z. Bisri, C. Piliego, J. Gao, and M. A. Loi, *Adv. Mater.* **2014**, *26*, 1176.

[51] M. A. B. H. Susan, T. Kaketo, A. Noda, and M. Watanabe, *J. Am. Chem. Soc.*, **2005**, *127*, 4976.

[52] J. Ho Cho, J. Lee, Y. Xia, B. Kim, Y. He, M. J. Renn, T. P. Lodge, and C. D. Frisbie, *Nat. Mater.*, **2008**, *7*, 900.

[53] F. Cicoira and C. Santato, *Adv. Funct. Mater.* **2007**, *17*, 3421.

[54] M. B. Smith and J. Michl, *Chem. Rev.* **2010**, *110*, 6891.

[55] S. Varghese and S. Das, *J. Phys. Chem. Lett.* **2011**, *2*, 863.

[56] W. Wang, T. Lin, M. Wang, T.-X. Liu, L. Ren, D. Chen, and S. Huang, *J. Phys. Chem. B*, **2010**, *114*, 5983.

[57] J. Luo, Z. Xie, J. W. Y. Lam, L. Cheng, H. Chen, C. Qiu, H. S. Kwok, X. Zhan, Y. Liu, D. Zhu, and B. Z. Tang, *Chem. Commun.* **2001**, **1740**.

[58] R. Katoh, K. Suzuki, A. Furube, M. Kotani, and K. Tokumaru, *J. Phys. Chem. C*, **2009**, *113*, 2961.

[59] D. J. Gundlach, J. A. Nichols, L. Zhou, and T. N. Jackson, *Appl. Phys. Lett.* **2002**, *80*, 2925.

[60] S. H. Lim, T. G. Bjorklund, F. C. Spano, and C. J. Bardeen, *Phys. Rev. Lett.* **2004**, *92*, 107402.

[61] E. Menard, V. Podzorov, S. H. Hur, A. Gaur, M. E. Gershenson, and J. A. Rogers, *Adv. Mater.* **2004**, *16*, 2097.

[62] H. Klauk, M. Halik, U. Zschieschang, G. Schmid, W. Radlik, and W. Weber, *J. Appl. Phys.* **2002**, *92*, 5259.

[63] W. Wu, Y. Liu, and D. Zhu, *Chem. Soc. Rev.* **2010**, *39*, 1489.

[64] M. J. Kang, I. Doi, H. Mori, E. Miyazaki, K. Takimiya, M. Ikeda, and H. Kuwabara, *Adv. Mater.* **2011**, *23*, 1222.

[65] A. N. Sokolov, S. Atahan-Evrenk, R. Mondal, H. B. Akkerman, R. S. Sanchez-Carrera, S. Granados-Focil, J. Schrier, S. C. B. Mannsfeld, A. P. Zoombelt, Z. N. Bao, and A. Aspuru-Guzik, *Nat. Commun.* **2011**, *2*, 437.

[66] R. D. McCullough, *Adv. Mater.* **1998**, *10*, 93.

[67] H. Saadeh, T. Goodson, III, and L. Yu *Macromolecules* **1997**, *30*, 4608.

[68] I. F. Perepichka, D. F. Perepichka, H. Meng, and F. Wudl, *Adv. Mater.* **2005**, *17*, 2281.

[69] A. Dadvand, A. G. Moiseev, K. Sawabe, W. H. Sun, B. Djukic, I. Chung, T. Takenobu, F. Rosei, and D. F. Perepichka, *Angew. Chem. Int. Ed.* **2012**, *51*, 3837.

[70] C. Di, F. Zhang, and D. Zhu, *Adv. Mater.* **2013**, *25*, 313.

[71] X. Guo, F. S. Kim, M. J. Seger, S. A. Jenekhe, and M. D. Watson, *Chem. Mater.* **2012**, *24*, 1434.

[72] C. Gu, W. Hu, J. Yao, and H. Fu, *Chem. Mater.* **2013**, *25*, 2178.

[73] J. Lee, A.-R. Han, J. Kim, Y. Kim, J. H. Oh, and C. Yang, *J. Am. Chem. Soc.* **2012**, *134*, 20713.

[74] J. Lee, A. R. Han, H. Yu, T. J. Shin, C. Yang, and J. H. Oh, *J. Am. Chem. Soc.* **2013**, *135*, 9540.

[75] T. Takenobu, S. Z. Bisri, T. Takahashi, M. Yahiro, C. Adachi, and Y. Iwasa, *Phys. Rev. Lett.* **2008**, *100*, 066601.

[76] M. A. Baldo, R. J. Holmes, and S. R. Forrest, *Phys. Rev. B* **2002**, *66*, 035321.

[77] F. Dinelli, R. Capelli, M. A. Loi, M. Murgia, M. Muccini, A. Facchetti, and T. J. Marks, *Adv. Mater.* **2006**, *18*, 1416.

[78] M. Melucci, L. Favaretto, M. Zambianchi, M. Durso, M. Gazzano, A. Zanelli, M. Monari, M. G. Lobello, F. De Angelis, V. Biondo, G. Generali, S. Troisi, W. Koopman, S. Toffanin, R. Capelli, and M. Muccini, *Chem. Mater.* **2013**, *25*, 668.

[79] H. Nakanotani, M. Saito, H. Nakamura, and C. Adachi, *Appl. Phys. Lett.* **2009**, *95*, 103307.

[80] K. Yamane, H. Yanagi, A. Sawamoto, and S. Hotta, *Appl. Phys. Lett.* **2007**, *90*, 162108.

[81] J. Swensen, C. Soci, and A. Heeger, *Appl. Phys. Lett.* **2005**, *87*, 253511.

[82] C. L. Donley, J. Zaumseil, J. W. Andreasen, M. M. Nielsen, H. Sirringhaus, R. H. Friend, and J. S. Kim, *J. Am. Chem. Soc.* **2005**, *127*, 12890.

[83] J. S. Kim, P. K. H. Ho, C. E. Murphy, and R. H. Friend, *Macromolecules* **2004**, *37*, 2861.

[84] X. Cheng, Y.-Y. Noh, J. Wang, M. Tello, J. Frisch, R.-P. Blum, A. Vollmer, J. P. Rabe, N. Koch, and H. Sirringhaus, *Adv. Funct. Mater.* **2009**, *19*, 2407.

[85] R. J. Kline, M. D. McGehee, E. N. Kadnikova, J. S. Liu, J. M. J. Frechet, and M. F. Toney, *Macromolecules* **2005**, *38*, 3312.

[86] E. Lunedei, P. Moretti, M. Murgia, M. Muccini, F. Biscarini, and C. Taliani, *Synth. Met.* **1999**, *101*, 592.

[87] J. Zaumseil, R. J. Kline, and H. Sirringhaus, *Appl. Phys. Lett.* **2008**, *92*, 073304.

[88] I. D. W. Samuel and G. A. Turnbull, *Chem. Rev.* **2007**, *107*, 1272.

[89] A. Yarif and P. Yeh, *Optical Waves in Crystals* (Wiley, New York, **2003**).

[90] L. Burgi, T. J. Richards, R. H. Friend, and H. Sirringhaus, *J. Appl. Phys.* **2003**, *94*, 6129.

[91] T. J. Richards and H. Sirringhaus, *J. Appl. Phys.* **2007**, *102*, 094510.

[92] M. C. Gwinner, S. Khodabakhsh, H. Giessen, and H. Sirringhaus, *Chem. Mater.* **2009**, *21*, 4425.

[93] M. D. McGehee, R. Gupta, S. Veenstra, E. K. Miller, M. A. Diaz-Garcia, and A. Heeger, *J. Phys. Rev. B* **1998**, *58*, 7035.

[94] R. Xia, G. Heliotis, Y. Hou, and D. D. C. Bradley, *Org. Electron.* **2003**, *4*, 165.

[95] C. M. Aguirre, C. Ternon, M. Paillet, P. Desjardins, and R. Martel, *Carbon Nano Lett.* **2009**, *9*, 1457.

[96] F. Cicoira, C. M. Aguirre, and R. Martel, *ACS Nano* **2011**, *5*, 283.

[97] N. Komatsu and F. Wang, *Materials* **2010**, *3*, 3818.

[98] M. C. Hersam, *Nat. Nanotechnol.* **2008**, *3*, 387.

[99] A. Nish, J. Y. Hwang, J. Doig, and R. J. Nicholas, *Nat. Nanotechnol.* **2007**, *2*, 640.

[100] J. Gao, M. A. Loi, E. J. F. de Carvalho, and M. C. dos Santos, *ACS Nano* **2011**, *5*, 3993.

[101] W. Gomulya, G. D. Costanzo, E. J. F. de Carvalho, S. Z. Bisri, V. Derenskyi, M. Fritsch, N. Fröhlich, S. Allard, P. Gordiichuk, A. Herrmann, S. J. Marrink, M. C. dos Santos, U. Scherf, and M. A. Loi, *Adv. Mater.* **2013**, *25*, 2948.

[102] M. C. Gwinner, F. Jakubka, F. Gannott, H. Sirringhaus, and J. Zaumseil, *ACS Nano* **2012**, *6*, 539.

[103] A. Nish, J. Y. Hwang, J. Doig, and R. J. Nicholas, *Nanotechnology* **2008**, *19*, 095603.

[104] J. Zaumseil, C. R. McNeill, M. Bird, D. L. Smith, P. P. Ruden, M. Roberts, M. J. McKiernan, R. H. Friend, and H. Sirringhaus, *J. Appl. Phys.* **2008**, *103*, 064517.

[105] T. Sakanoue, M. Yahiro, C. Adachi, H. Uchiuzou, T. Takahashi, and A. Toshimitsu, *Appl. Phys. Lett.* **2007**, *90*, 171118.

[106] M. Melucci, M. Zambianchi, L. Favaretto, M. Gazzano, A. Zanelli, M. Monari, R. Capelli, S. Troisi, S. Toffanin, and M. Muccini, *Chem. Commun.* **2011**, *47*, 11840.

[107] A. Facchetti, M. H. Yoon, C. L. Stern, G. R. Hutchison, M. A. Ratner and T. J. Marks, *J. Am. Chem. Soc.* **2004**, *126*, 13480.

[108] A. Mishra, C. Q. Ma and P. Bauerle, *Chem. Rev.* **2009**, *109*, 1141.

[109] G. Barbarella, M. Melucci and G. Sotgiu, *Adv. Mater.* **2005**, *17*, 1581.

[110] L. Pauling, *The Nature of the Chemical Bond* (Cornell University, Ithaca, NY, **1945**).

[111] E. Tedesco, F. Della Sala, L. Favaretto, G. Barbarella, D. Albesa-José, D. Pisignano, G. Gigli, R. Gingolani, and K. D. M. Harris, *J. Am. Chem. Soc.* **2003**, *125*, 277.

[112] A. Facchetti, M. Mushrush, M.-H. Yoon, G. R. Hutchison, M. A. Ratner, and T. J. Marks, *J. Am. Chem. Soc.* **2004**, *126*, 13859.

[113] J. L. Bredas, D. Beljonne, J. Cornil, J. P. Calbert, Z. Shuai, and R. Silbey, *Synth. Met.* **2001**, *125*, 107.

[114] X. C. Li, H. Sirringhaus, F. Garnier, A. B. Holmes, S. C. Moratti, N. Feeder, W. Clegg, S. J. Teat, and R. H. Friend, *J. Am. Chem. Soc.* **1998**, *120*, 2206.

[115] M. Melucci, M. Durso, C. Bettini, M. Gazzano, L. Maini, S. Toffanin, S. Cavallini, M. Cavallini, D. Gentili, V. Biondo, G. Generali, F. Gallino, R. Capelli, and M. Muccini, *J. Mater. Chem. C*, **2014**, *2*, 3448.

[116] S. M. Sze, *Physics of Semiconductor Devices* (Wiley, New York, **1981**), 245.

[117] J. Campbell Scott, *J. Vac. Sci. Technol., A* **2003**, *21*, 521.

[118] M. Schidleja, C. Melzer, and H. von Seggern, *Adv. Mater.* **2009**, *21*, 1172.

[119] R. A. Street and A. Salleo, *Appl. Phys. Lett.* **2002**, *81*, 2887.

[120] P. V. Pesavento, *et al.*, *J. Appl. Phys.* **2004**, *96*, 7312.

[121] C. D. Dimitrakopolous, A. R. Brown, and A. Pomp, *J. Appl. Phys.* **1996**, *80*, 2501.

[122] I. Kymissis, C. D. Dimitrakopolous, and S. Purushothaman, *IEEE Trans. Elec. Dev.* **2001**, *48*, 1060.

[123] Z. B. J. Locklin *Organic Field-effect Transistors*, (CRC Group Taylor and Francis Group, Boca Raton, FL, **2007**).

[124] R. Schmechel, M. Ahles, and H. von Seggern, *J. Appl. Phys.* **2005**, *98*, 084511.

[125] M. Schidleja, C. Melzer, and H. von Seggern, *Appl. Phys. Lett.* **2009**, *94*, 123307.

[126] C. Rost, S. Karg, W. Riess, M. A. Loi, M. Murgia, and M. Muccini, *Synth. Met.* **2004**, *146*, 237.

[127] J. Reynaert, D. Cheyns, D. Janssen, R. Muller, V. I. Arkhipov, J. Genoe, G. Borghs, and P. Heremans, *J. Appl. Phys.* **2005**, *97*, 114501.

[128] J. Swensen, D. Moses, and A. J. Heeger, *Synth. Met.* **2005**, *153*, 53.

[129] K. Walzer, B. Maennig, M. Pfeiffer, and K. Leo, *Chem. Rev.* **2007**, *107*, 1233.

[130] A. Nollau, M. Pfeiffer, T. Fritz, and K. Leo, *J. Appl. Phys.* **2000**, *87*, 4340.

[131] J. Wang, J. Chang, and H. Sirringhaus, *Appl. Phys. Lett.* **2005**, *87*, 3.

[132] I. H. Campbell, S. Rubin, T. A. Zawodzinski, J. D. Kress, R. L. Martin, D. L. Smith, N. N. Barashkov, and J. P. Ferraris, *Phys. Rev. B* **1996**, *54*, 14321.

[133] I. H. Campbell, J. D. Kress, R. L. Martin, D. L. Smith, N. N. Barashkov, and J. P. Ferraris, *Appl. Phys. Lett.* **1997**, *71*, 3528.

[134] B. de Boer, A. Hadipour, M. M. Mandoc, T. van Woudenbergh, and P. W. M. Blom, *Adv. Mater.* **2005**, *17*, 621.

[135] N. Koch, *Chem. Phys. Chem.* **2007**, *8*, 1438.

[136] S. Braun and W. R. Salaneck, *Chem. Phys. Lett.* **2007**, *438*, 259.

[137] H. Sirringhaus, R. J. Wilson, R. H. Friend, M. Inbasekaran, W. Wu, E. P. Woo, M. Grell, and D. D. C. Bradley, *Appl. Phys. Lett.* **2000**, *77*, 406.

[138] J. Zaumseil, C. Groves, J. M. Winfield, N. C. Greenham, and H. Sirringhaus, *Adv. Funct. Mater.* **2008**, *18*, 3630.

[139] H. J. Bolink, E. Coronado, J. Orozco, and M. Sessolo, *Adv. Mater.* **2009**, *21*, 79.

[140] D. Kabra, M. H. Song, B. Wenger, R. H. Friend, and H. J. Snaith, *Adv. Mater.* **2008**, *20*, 3447.

[141] N. Tokmoldin, N. Griffiths, D. D. C. Bradley, and S. A. Haque, *Adv. Mater.* **2009**, *21*, 3475.

[142] B. S. Ong, C. Li, Y. Li, Y. Wu, and R. Loutfy, *J. Am. Chem. Soc.* **2007**, *129*, 2750.

[143] G. Adamopoulos, A. Bashir, P. H. Wöbkenberg, D. D. C. Bradley, and T. D. Anthopoulos, *Appl. Phys. Lett.* **2009**, *95*, 133507.

[144] M. C. Gwinner, Y. Vaynzof, K. K. Banger, P. K. H. Ho, R. H. Friend, and H. Sirringhaus, *Adv. Funct. Mater.* **2010**, *20*, 3457.

[145] S. K. Hau, H.-L. Yip, H. Ma, and A. K.-Y. Jen, *Appl. Phys. Lett.* **2008**, *93*, 233304.

[146] Y. Vaynzof, D. Kabra, L. Zhao, P. K. H. Ho, A. T.-S. Wee, and R. H. Friend, *Appl. Phys. Lett.* **2010**, *97*, 033309.

[147] V. I. Arkhipov, U. Wolf, and H. Bässler, *Phys. Rev. B* **1999**, *59*, 7514.

5

CHARGE-TRANSPORT AND PHOTOPHYSICAL PROCESSES IN OLETs

In this chapter, the charge-transport properties, excitonic processes, and emitting characteristics in organic light-emitting transistors (OLETs) with different geometries are discussed. The analysis of the charge-transport fundamentals relies on theories commonly used for organic field-effect transistors, with emphasis on ambipolar devices, which is particularly relevant for OLETs. The key fundamental excitonic processes to be considered for a description of the optoelectronic properties of OLETs are introduced, together with the models applied so far to describe the OLET behavior. One of the most distinguishing features of OLETs, that is, the emitting area within the transistor channel, is analyzed for different classes of materials and for different structures of the active channel.

5.1 CHARGE TRANSPORT IN OLETs

The exact nature of charge transport in organic semiconductors is still open to debate. Nevertheless, one can make a clear distinction between disordered semiconductors such as amorphous polymers and highly ordered organic single crystals, at the opposite ends of the spectrum. Charge transport in disordered semiconductors is generally described by thermally activated hopping of charges through a distribution of localized states or shallow traps. Bassler [1]

Organic Light-Emitting Transistors: Towards the Next Generation Display Technology,
First Edition. Michele Muccini and Stefano Toffanin.
© 2016 John Wiley & Sons, Inc. Published 2016 by John Wiley & Sons, Inc.

has described this density of states (DOS) as a Gaussian distribution in order to model charge transport in time-of-flight experiments.

The width of the Gaussian DOS is defined by the spatial and energetic disorder within the semiconductor and can be determined by temperature-dependent mobility measurements [2]. A broader DOS leads to lower mobilities and a stronger temperature dependence.

A variable-range hopping model, where charges can hop a short distance with a high activation energy or a long distance with a low activation energy, was used by Vissenberg and Matters [3]. They further assumed an exponential distribution of localized states, which represents the tail of a Gaussian DOS that dominates the transport characteristics at low carrier concentrations. The Vissenberg–Matters model predicts an increase of the field-effect mobility with increasing gate voltage given that the accumulated charge carriers fill the lower lying states of the organic semiconductor first, and any additional charges in the accumulation layer will occupy states at relatively high energies. Thus, additional charges will require a lower activation energy to hop between sites. This dependence of the mobility on charge density, and thus gate voltage, has been observed for many disordered semiconductors, and the Vissenberg–Matters model proved to be very useful to model organic field-effect transistors (OFETs) and reconcile charge mobilities in organic diode structures and field-effect transistors [4–6].

Note that the model does not make any assumptions on the transport mechanism. The main results are very similar to those of the hopping model; that is, the mobility is thermally activated and follows a power-law dependence with the gate voltage. The main interest of the model is thus to explain the behavior found in most OFET devices made of polycrystalline films of small conjugated molecules. However, many organic systems present transport properties, which cannot be thoroughly explained within the Vissenberg–Matters model. For example [7], the temperature-independent mobility reported for pentacene [8] and oligothiophene [9,10] thin-film transistors is not accounted for.

Moreover, several high-mobility polycrystalline conjugated polymers have been reported to exhibit highly nonlinear transport properties at low temperatures [11,12]. These have been interpreted as a manifestation of one-dimensional Luttinger liquid physics [13], but their microscopic origin remains unclear [14,15]. Indeed, it is worth noting that the apparent band-like temperature dependence of the field-effect mobility of solution-processed pentacene derivative was ascribed to localized transport limited by thermal diffusion and not to extended-state conduction [16].

In the case of highly ordered molecular crystals, such as rubrene, tetracene, and pentacene, experimental data seem to rule out the transport to occur through hopping, at least at low temperature. Temperature-dependent time-of-flight [17], time-resolved terahertz spectroscopy [18], and field-effect transistor [19] measurements on high-purity crystals [20] suggest band-like transport in delocalized states instead of hopping transport. Concomitantly, the mean free path of charge carriers at high temperatures (above 150 K) is found to be comparable with the crystal unit cell, which is not coherent with a diffusion-limited transport [21]. Recent theoretical studies suggest that thermal motion modulates the intermolecular electronic coupling (transfer integrals) between molecules in organic crystals due to their weak interaction, which could lead to localization of charge carriers even in highly ordered systems [22]. Furthermore, the polarizability of the gate dielectric can cause localization in organic single-crystal field-effect transistors, as shown by Hulea *et al.* [23].

However, in spite of recent advances [4–6,17], a convincing and comprehensive modeling of the process of charge-carrier transport in molecular crystals is not available yet [2,24].

A common building block for the modeling of disordered semiconductors and highly ordered organic single crystals is the transfer integral representing the electronic coupling of adjacent molecules and the polaronic relaxation energy, which is the energy gained when a charge geometrically relaxes over a single molecule or polymer segment. This is an important parameter, which determines the probability of charge transport from one molecule to another and depends strongly on the particular molecule and the relative position of the interacting units [25].

The simplest treatment of the current–voltage characteristics of organic ambipolar light-emitting transistors is based on the macroscopic standard field-effect transistor equations. Within the gradual channel approximation, and considering charge-density-independent mobility and ohmic injecting contacts [26], the current continuity equations can be reduced to

$$I_{ds} = \frac{W}{L} \mu C_i \left[\left(V_{gs} - V_t \right) V_{ds} - \frac{V_{ds}^2}{2} \right] \tag{5.1}$$

for V_{gs} in accumulation regime and $|V_{ds}| \le |V_{gs}|$

$$I_{ds,\,sat} = \frac{W}{L} \mu C_i \left[V_{gs} - V_t \right]^2 \tag{5.2}$$

If V_{ds} is small, then the charge density across the channel is quasiconstant and the extracted mobility is well suitable for comparison with theoretical predictions. The most direct way to extract the mobility would be to solve the aforementioned equation for μ:

$$\mu = \begin{cases} \dfrac{I_{ds}}{\dfrac{W}{L} C_i \left[\left(V_{gs} - V_t \right) V_{ds} - \dfrac{V_{ds}^2}{2} \right]} & \text{Linear regime} \\[6ex] \dfrac{I_{ds,\,sat}}{\dfrac{W}{L} C_i \left[V_{gs} - V_t \right]^2} & \text{Saturation regime} \end{cases} \qquad (5.3)$$

It is common to use the derivative of the aforementioned equation in order to extract the mobility [27,28] as a function of gate voltage.

However, if the mobility is density (gate voltage) dependent, the method is not reliable and the formula may lead to large errors. To demonstrate this, one can write the derivative of the equation for the linear regime:

$$\frac{\partial}{\partial V_{gs}} I_{ds} = \frac{\partial \mu}{\partial V_{gs}} \left\{ \frac{W}{L} C_i \left[\left(V_{gs} - V_t \right) V_{ds} - \frac{V_{ds}^2}{2} \right] \right\} + \mu \left\{ \frac{W}{L} C_i V_{ds} \right\} \qquad (5.4)$$

Only when the mobility is independent of the gate bias (i.e., of the charge density), one can neglect the first term of Equation 5.4, which is then simplified to

$$\mu = \frac{\dfrac{\partial}{\partial V_{gs}} I_{ds}}{\left\{ \dfrac{W}{L} C_i V_{ds} \right\}} \qquad (5.5)$$

In Figure 5.1, the experimentally measured source–drain current in a polymer-based (MEH-PPV, poly[2-methoxy-5-(2′-ethylhexyloxy)-p-phenylene vinylene]) single-layer OFET [29] is reported, together with the mobility, as a function of gate voltage, which is derived from Equation 5.3 [30]. The mobility shows about an order of magnitude increase from 10^{-6} to 10^{-5} cm^2/V s [27].

To analyze the consequences of neglecting the gate voltage dependence of charge-carrier mobility, the first and second terms of Equation 5.4 are reported as a function of the gate voltage in Figure 5.2.

As it can be seen, the two terms in the equation can be quantitatively comparable, leading to potential errors in the order of 100%. In addition, Equation 5.5 also exhibits an incorrect gate voltage dependence.

Even in the case in which the voltage dependence would be correctly described, one would still need to determine the functional relation between charge-carrier density and applied gate voltage.

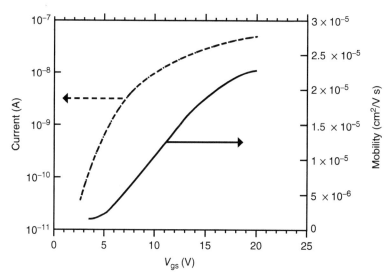

FIGURE 5.1 Drain–source current (dashed line) and the mobility (full line) deduced using Equation 5.3 as a function of the gate voltage. [*Source*: Roichmann *et al*. [30]. Adapted with permission from Wiley.]

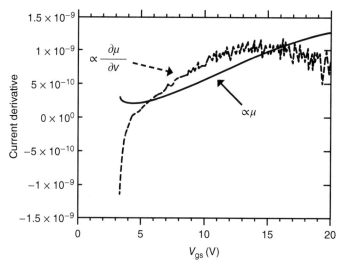

FIGURE 5.2 First (dashed line) and second (solid line) terms of Equation 5.4 implementing the data reported in Figure 5.1. [*Source*: Roichmann *et al*. [30]. Adapted with permission from Wiley.]

Roichmann *et al.* [30] proposed a model to reproduce qualitatively and quantitatively, within the same physical frame, the experimental outputs of different organic devices (organic light-emitting diode (OLED), OFET, organic photovoltaic (OPV)). In particular, they suggested that the extended Gaussian disorder model can self-consistently and qualitatively explain the operation of a range of device structures.

Specifically for OFETs, it is possible to derive the electric field E normal to the device plane and variable along the channel length (L) by assuming that the ratio between the charge-carrier diffusion (D) and mobility (μ) constants is invariant along the transistor channel length:

$$E = \sqrt{\frac{2aD}{\mu}} \tan\left[\sqrt{\frac{a\mu}{2D}}(x-L)\right] \tag{5.6}$$

with a being a numerical constant.

Thus, the charge-carrier density across the organic layer is given by

$$\rho = \frac{\varepsilon_r \varepsilon_0}{q}\frac{\partial E}{\partial x} = \frac{a\varepsilon_r \varepsilon_0}{q}\left\{\tan\left[\sqrt{\frac{a\mu}{2D}}(x-L)\right]^2 + 1\right\} \tag{5.7}$$

where ε_0 and ε_r are the permittivities of the vacuum and the organic material, respectively.

In principle, this formalism can be applied to DOS with any shapes, and optoelectronic devices with generic structure, provided that the unique properties of organic semiconductors are accounted for by appropriate expressions for mobility and/or diffusion coefficients. However, the analysis and modeling of polymer- or small-molecule-based devices are often performed using different models and sometimes even different physical pictures. Indeed, when compared to the tremendous progress that the performances of field-effect organic thin-film transistors have experienced during the past 10 years, the theory of charge transport has dramatically lagged behind.

In real organic transistors, charge transport is most often time-limited by localized states induced by defects and unwanted impurities, as witnessed by strongly sample-dependent performances. Two models are useful when dealing with such phenomena: the multiple trapping and thermal release (MTR) model and the variable range hopping (VRH) model. While hopping transport is appropriate to describe charge transport in disordered systems, the MTR model [31] applies to well-ordered materials, prototypes of which are vapor-deposited small molecules such as pentacene or oligothiophenes, where thermally activated mobility is often observed. The basic assumption of the model is a distribution of localized energy levels located in the vicinity

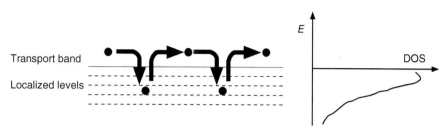

FIGURE 5.3   Principle of charge transport limited by multiple trapping and thermal release.

of the transport band edge. During their transit in the delocalized band, the charge carriers interact with the localized levels through trapping and thermal release (Figure 5.3).

The model is based on the following main assumptions: (i) carriers arriving at a trap are instantaneously captured with a probability close to 1; and (ii) the release of trapped carriers is controlled by a thermally activated process. The resulting effective mobility μ_{eff} is related to the mobility in the transport band μ_0 by

$$\mu_{\text{eff}} = \mu_0 \alpha e^{-(E_c - E_t)/kT} \tag{5.8}$$

where E_c is the energy of the transport band edge. In the case of a single trap level of energy E_t and DOS N_t, the total charge-carrier concentration splits into a concentration of free carriers $n_f = N_c e^{-(E_c - E_t)/kT}$, where N_c is the effective DOS at transport band edge, and a concentration of trapped carriers

$$n_t = N_t e^{-(E_t - E_F)/kT} \tag{5.9}$$

The ratio of trapped to total densities is given by

$$\Theta = \frac{n_t}{n_t + n_f} = \frac{1}{1 + \dfrac{N_t}{N_c} e^{(E_c - E_t)/kT}} \approx \frac{N_t}{N_c} e^{-(E_c - E_t)/kT} \tag{5.10}$$

In that instance, the effective mobility is $\mu_{\text{eff}} = \Theta\mu_0$, so that E_t in Equation 5.8 is the energy of the single trap level. If traps are energy distributed, distribution-dependent effective values of E_t and α must be estimated. In all circumstances, whichever the actual energy distribution of traps, the main feature predicted by the MTR model is thermally activated mobility.

An important outcome of the MTR model is that in the case of an energy-distributed DOS, mobility is gate-voltage dependent. The mechanism is schematically illustrated in Figure 5.4.

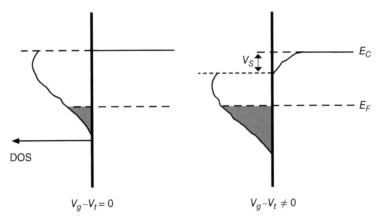

FIGURE 5.4 Gate-voltage-dependent mobility induced by an energy-distributed density of traps.

When a bias is applied to the gate, a potential V_{gs} develops at the insulator–semiconductor interface, which results in shifting by the same amount the Fermi level toward the transport band edge, thus partly filling the distribution of localized states. Accordingly, the energy distance between the filled traps and the transport band edge is reduced, so trapped-carrier release is made easier, and the effective mobility increases. Such a gate-voltage dependence of mobility has indeed been observed in several systems [32,33].

The gate voltage dependence is determined by the shape of the DOS. Due to disorder and variation of interaction energies, thin films of disordered semiconductors are characterized by a spatial and energetic spread of charge-transport sites described by a Gaussian density of state distribution [1]. Furthermore, for disordered small-molecule systems, the electrostatic field from a random distribution of static or induced dipoles leads to a Gaussian DOS function [34]. For a system both with negligible background doping and at typical gate-induced carrier densities, the carrier mobility resulting from hopping in a Gaussian DOS can be approximated by the mobility resulting from hopping in an exponential DOS [3,35,36].

The almost linear relationship (on a double-log scale) between the field-effect mobility and the carrier concentration at the gate dielectric/semiconductor interface [36], which is often observed, suggests that an exponential DOS is a good approximation for the DOS in the case of intermediate and high carrier concentrations [37].

Thus, the trap distribution used in the MTR model is also an exponential band tail. The exponential distribution leads to an analytical form of the gate voltage dependence of the mobility. A generic exponential distribution of traps is given by

$$N_t(E) = \frac{N_{t0}}{kT_0} e^{-(E_c - E)/kT_0} \tag{5.11}$$

where N_{t0} is the total density (per unit area) of traps, k is the Boltzmann constant, and T_0 is a characteristic temperature that accounts for the slope of the distribution. The trapped charge amount is connected to the density of traps through

$$n_t = q \int_{-\infty}^{+\infty} N_t(E) f(E) dE \tag{5.12}$$

where $f(E)$ is the Fermi distribution. If $N_t(E)$ is a slowly varying function, the Fermi distribution can be approximated to a step function; that is, it equals 0 for $E < E_F$ and 1 for $E > E_F$. The integration of Equation 5.12 leads to

$$n_t \approx q N_t (E_{F0} + qV_s) = n_{t0} e^{qV_s/kT_0} \tag{5.13}$$

As stated earlier, we have made use of the fact that the Fermi level E_F is shifted toward the band edge E_C from the value at zero gate bias by an amount qV_s (Figure 5.4). $n_{t0} = N_{t0} e^{-(E_c - E_{F0})/kT_0}$ is the density of trapped charge at zero gate voltage. Assuming that $n_f \ll n_t$, we finally obtain [34]

$$\mu_{\text{eff}} = \mu_0 \frac{N_c}{N_{t0}} \left\{ \frac{C_i (V_g - V_t)}{q N_{t0}} \right\}^{\frac{T_0}{T} - 1} \tag{5.14}$$

Note that the gate voltage dependence has the form of a power law in $V_g - V_t$ [32]. In single-crystal transistors, the mobility is typically gate-voltage independent [38], which confirms that the V_g dependence originates from localized levels associated with chemical and physical defects.

In the case of the variable-range hopping (VRH) model, the field-effect transport in amorphous organic transistors can be derived by considering the charge transport governed by hopping, that is, the thermally activated tunneling of carriers between localized states, rather than by the activation of carriers to a transport level. A carrier may either hop over a small distance with a high activation energy or hop over a long distance with a low activation energy. The temperature dependence of the carrier transport in such a system is strongly dependent on the density of localized states. As we have already stated, an applied gate voltage in a field-effect transistor gives rise to the accumulation of charge in the region of the semiconducting layer that is close to the insulator. As these accumulated charge carriers fill the lower lying states of the organic semiconductor, any additional charges in the accumulation layer will occupy states at higher energies. Consequently, these additional charges will—on average—require less activation energy to hop away to a neighboring site. This results in a higher mobility with increasing gate voltage.

To gain a deeper insight into the physical meaning of this model, let us start by deriving an expression for the conductivity as a function of temperature T and charge-carrier density of a VHR system with exponential distribution of localized-state energies. At low carrier densities and low T, the tail of the density of (localized) states determines the transport properties. In the model implemented by Vissenberg and Matters [3], the exponential DOS function is defined as

$$g(E) = \frac{N_t}{kT_0} \exp\left(\frac{E}{kT_0}\right) \quad (-\infty < E \leq 0) \tag{5.15}$$

where N_t is the number of states per unit volume, k is the Boltzmann's constant, and T_0 is a parameter that indicates the width of the exponential distribution. We take $g(E) = 0$ for positive values of E. The results are not expected to be qualitatively different for a different choice of $g(E)$, as long as $g(E)$ increases strongly with E.

Let us consider the system filled with charge carriers, such that a fraction $\delta[0,1]$ of the localized states is occupied by a carrier such that the density of charge carriers is δN_t. In equilibrium, the energy distribution of the carriers is given by the Fermi–Dirac distribution $f(E,E_F)$, where E_F is the Fermi energy. For a given carrier occupation δ, the position of the Fermi E_F is fixed by the condition:

$$\delta = \frac{1}{N_t} \int dE \, g(E) f(E_i, E_F)$$

$$\approx \exp\left(\frac{E_F}{kT_0}\right) \Gamma(1 - T/T_0) \Gamma(1 + T/T_0) \tag{5.16}$$

where $\Gamma(z) = \int_0^\infty dy \exp(-y) \, y^{z-1}$. The assumption that most carriers occupy the sites with energies $E << 0$ (i.e., $-E_F >> kT_0$) is used. This condition is fulfilled when δ and T are low and breaks down at $T \geq T_0$ (where $\Gamma(1 - T/T_0)$ diverges). At such temperatures, the assumption that only the tail of the DOS is important no longer holds, as the majority of the carriers is located close to $E = 0$.

The transport process is governed by the hopping of carriers between localized states, which is strongly dependent on the hopping distances as well as the energy distribution of the states. According to percolation theory [39,40], the conductivity of the system is given by

$$\sigma = \sigma_0 e^{-s_c} \tag{5.17}$$

where σ_0 is the (unknown) prefactor for conductivity and s_c is the exponent of the percolation conductance $G_C = G_0 \exp[-s_c]$ at the onset of the percolation.

The conductivity as a function of the occupation δ and the temperature T is given by

$$\sigma(\delta, T) = \sigma_0 \left(\frac{\pi N_t \delta (T_0 / T)^3}{(2\alpha)^3 B_c \Gamma(1 - T / T_0) \Gamma(1 + T / T_0)} \right)^{T_0 / T} \tag{5.18}$$

where α is the effective overlap parameter and B_c the critical number for the onset percolation:

$$B_c \approx \pi \left(\frac{T_0}{2\alpha T} \right)^3 N_t \exp\left(\frac{E_F + s_c kT}{kT_0} \right) \tag{5.19}$$

Note that the temperature dependence of the conductivity is Arrhenius-like $\sigma \sim \exp[-E_a/(k_B T)]$ with an activation energy E_a that is weakly (logarithmically) temperature dependent. The conductivity increases superlinearly with the density of carriers. Indeed, an increase in the carrier density gives rise to an increase in the average energy, thus facilitating an activated jump to a specific transport energy level.

As we have discussed previously for the MTR model, a field-effect mobility can be derived from the conductivity by considering that the occupation δ at a distance x from the dielectric interface is given by

$$\delta(x) = \delta_0 \exp\left(\frac{eV(x)}{kT_0} \right) \tag{5.20}$$

where $V(x)$ is the potential at position x, and δ_0 is the carrier occupation far from the semiconductor–insulator interface where $V(x) = 0$.

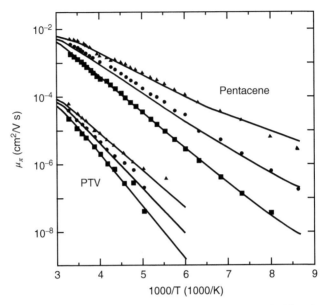

FIGURE 5.5 Field-effect mobility in a pentacene and a polythienylene vinylene (PTV) thin-film transistor as a function of the temperature T for different gate voltages $V_G = 220\,\text{V}$ (triangles), $210\,\text{V}$ (circles), and $25\,\text{V}$ (squares). The solid lines are according to Equation 5.21. [*Source*: Vissenberg *et al.* [3], Figure 1. Adapted with permission of American Physical Society.]

Thus, implementing the transconductance and the source–drain current in the linear-regime formulas in FET theory that we previously derived (Eq. 5.3), the field-effect mobility μ can be expressed as

$$\mu = \frac{\sigma_0}{e}\left(\frac{\pi\left(T_0 / T\right)^3}{\left(2\alpha\right)^3 B_c \Gamma\left(1 - T / T_0\right)\Gamma\left(1 + T / T_0\right)}\right)^{T_0/T}$$

$$x = \left[\frac{\left(C_i V_{gs}\right)^2}{2kT_0\varepsilon_r}\right]^{T_0/T} \tag{5.21}$$

where it is assumed that the thickness t of the semiconductor layer is sufficient to have $V(t) = 0$. The capacitance of the insulator is C_i and the dielectric constant of the semiconductor is ε_r.

These results were first used by Vissenberg and Matters [3] to interpret the experimentally observed temperature and gate-voltage dependence of the field-effect mobility in both pentacene and polythienylene vinylene (PTV) organic thin-field transistors (Figure 5.5).

The main difference between pentacene and PTV appears to be in the overlap parameter α, which determines the tunneling process between different sites. We note that this key parameter is absent in a multiple-trapping model, where the transport is governed by thermal activation from traps to a conduction band and subsequent detrapping, without involving a tunneling step. As the length scale α^{-1} is smaller than the size of a molecule, one must be cautious not to interpret α^{-1} simply as the decay length of the electronic wave function. The size and shape of the molecules and the morphology of the organic film are also expected to have an important influence on the tunneling probability.

The observed difference in α^{-1} may be due to the fact that there is more steric hindrance in the polymer PTV than in small molecular systems. The better stacking properties of pentacene give rise to a larger area of overlap of the electronic wavefunctions, which results in a larger effective overlap α^{-1} in the model.

Thus, the VRH model gives an analytical description for the bulk conductivity in the case of disordered organic semiconductors as a function of carrier density in the linear regime ($V_{ds} \rightarrow 0$) where the potential in the channel is a good approximation uniform between source and drain [41].

It has been shown that the formalism can be applied to a wide range of bias conditions, both unipolar and ambipolar OFETs, which is a necessary requirement for implementing the model in the description of the light-emitting transistors. In particular, Smits et al. [37] used the VRH model for describing the electrical characteristics of ambipolar organic transistors based on narrow-gap organic molecule. A schematic layout of the model bottom gate and bottom contact transistor is reported in Figure 5.6a, where as usual L is the channel length, W is the channel width in the y direction, and t is the semiconductor thickness. The x direction is the direction between source and drain, and the z direction is the direction perpendicular to the channel.

Field-effect transistors are fabricated on a heavily doped n-silicon wafer acting as the gate electrode with a 200 nm thick layer of thermally grown silicon dioxide as gate dielectric.

A 200 nm thick layer of thermally grown silicon dioxide is used as gate dielectric whose surface is subsequently passivated by a hexamethyldisilazane (HMDS) treatment. As an ambipolar semiconductor, a thin film of small-band-gap organic molecule, nickel-dithiolene (NiDT) (Figure 5.6a), was spin-coated onto the SiO_2-functionalized surface.

Gold is used as electrode metal for both source and drain contacts since the Fermi level of gold is in between the highest occupied molecular orbital (HOMO) and lowest unoccupied molecular orbital (LUMO) energy levels of

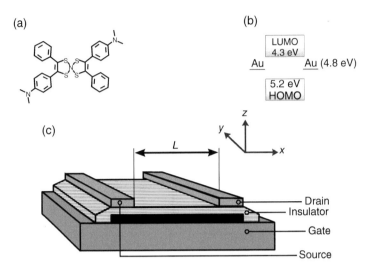

FIGURE 5.6 (a) Molecular structure of nickel dithiolene (bis[4-dimethylaminodi-thiobenzyl]-nickel). (b) Band diagram of nickel dithiolene. (c) Schematic of a bottom gate and contact transistor used as experimental model system. The x, y, and z directions are indicated to visualize the directions referred to in the text.

NiDT (Figure 5.6b). This implies that the barrier for injection of holes in the HOMO and for electrons in the LUMO is expected to be smaller than 0.5 eV, which is supposed to enable in principle injection of both types of charges in the field-effect transistors.

To derive the source–drain current (I_{ds}) as a function of the gate bias, source and drain contact resistances are considered negligible, and gradual channel approximation is used so that charge transport can be treated independently along the directions normal to the substrate and parallel to the channel length.

Firstly, the sheet conductance in the device channel is defined as a function of only the effective gate potential and temperature by integrating the conductivity along the thickness of the active layer (z-direction) with the assumption that the carrier density at the top of the semiconductor is negligible. Thus, the source–drain current can be calculated by integrating the sheet conductance along the channel length as

$$I_{ds} = \gamma \frac{W}{L} \frac{T}{2T_0} \frac{T}{2T_0 - T} \left[\left(V_{gs} - V_t \right)^{2T_0/T} - \left(V_{gs} - V_t - V_{ds} \right)^{2T_0/T} \right] \quad (5.22)$$

for $V_{gs} - V_t \geq V_{ds}$ with

$$\gamma = \frac{\sigma_0}{q} \left(\frac{\left(\frac{T_0}{T} \right)^4 \sin\left(\pi \frac{T}{T_0} \right)}{(2\alpha)^3 B_c} \right)^{T_0/T} \left(\frac{1}{2kT_0 \varepsilon_r \varepsilon_0} \right)^{(T_0/T)-1} C_i^{(2T_0/T)-1} \qquad (5.23)$$

When $V_{gs} - V_t < V_{ds}$ (so that the transistor is operated in saturation regime), the source current can be directly derived from Equation 5.22 by substituting V_{ds} with $V_{gs} - V_t$ under the assumption that the shortening of the channel length due to the *pinch-off* occurrence is negligible.

$$I_{ds}^{sat} = I_s \left(V_{ds} = V_{gs} - V_t \right) = \gamma \frac{W}{L} \frac{T}{2T_0} \frac{T}{2T_0 - T} \left(V_{gs} - V_t \right)^{2T_0/T} \qquad (5.24)$$

As expected, in the limit of small drain bias (V_{ds}), I_{ds} is a linear function of V_{ds} and Equation 5.22 is consistent with the results for the linear regime reported by Vissenberg *et al.* Indeed, from the analytical expression of Equation 5.22, it is possible to derive the model for the ambipolar transport in field-effect devices based on organic thin films. Ambipolar regime can occur when $V_{ds} > V_{gs} - V_t > 0$ and $V_{ds} < V_{gs} - V_t < 0$ and is characterized by a hole and an electron accumulation layer next to the respective electrodes that meet at some point within the transistor channel. At a first approximation, it can be considered that a narrow transition region, which acts as a pn junction, separates the accumulation regions.

Within this model based on VRH and percolation, the bimolecular recombination of holes and electrons with an infinite rate constant is assumed. Thus, charge-carrier density profiles are supposed not to interpenetrate in the channel, but to meet at a single plane $x = x_0$, where the effective gate potential is zero. As we discuss in Section 5.3, this assumption is related to unphysical determination of the width of the emission zone in OLETs within this model.

Thus, imposing that both currents are equal because recombination rate is infinite, the position x_0 of the pn junction is defined as

$$x_0 = \frac{\left(V_{gs} - V_t \right)^{\left(2T_{0,e}/T \right)^{-1}}}{\left(V_{gs} - V_t \right)^{\left(2T_{0,e}/T \right)^{-1}} + a \left(V_{ds} - V_{gs} - V_t \right)^{\left(2T_{0,h}/T \right)^{-1}}} L \qquad (5.25)$$

with

$$a = \frac{\left(2T_{0,e} - T\right)2T_{0,e}}{\left(2T_{0,h} - T\right)2T_{0,h}} \frac{\gamma_h}{\gamma_e} \qquad (5.26)$$

The expression for the source–drain current in the ambipolar regime can be obtained as

$$I_{ds} = \frac{W}{L}\left(\gamma_e \frac{T}{2T_{0,e}} \frac{T}{2T_{0,e} - T}\left(V_{gs} - V_t\right)^{\left(2T_{0,e}/T\right)} \right.$$
$$\left. + \gamma_h \frac{T}{2T_{0,h}} \frac{T}{2T_{0,h} - T}\left(-V_{gs} + V_t + V_{ds}\right)^{\left(2T_{0,e}/T\right)} \right) \qquad (5.27)$$

Surprisingly, the current calculated in this way is exactly equal to the sum of the electron and hole currents that would be present in the absence of a current of the other polarity: the ambipolar transistor is then represented by a unipolar n-type transistor in parallel with a unipolar p-type transistor, both operating in saturation (see Eq. 5.24). The hole and electron currents are constant, but not necessarily equal, throughout the whole channel.

In order to implement input parameters for the model, the authors determined independently the value for the threshold voltage from metal–insulator–semiconductor diode measurements. Indeed, the model does not require the explicit definition of the threshold and saturation voltages as input parameters, which are rather ambiguously defined.

The other transport process parameters, namely the width of the exponential DOS, the wavefunction overlap localization length, and the conductivity prefactor for holes and electrons, are determined by the unipolar response of the transistors. Using these parameters, an excellent agreement between measured and calculated currents is obtained for the ambipolar transfer curves as a function of temperature (Figure 5.7).

Despite this agreement, the output of the model that the electrons and holes move independently without forming of a pn junction is to be considered as unphysical. Moreover, it has to be noted that with this assumption, the recombination rate cannot be determined from the experimental current–voltage characteristics.

The exciton recombination width and light-emitting area in OLETs are the subject of extensive discussion in Section 5.3.

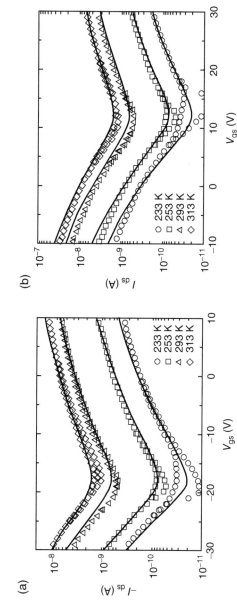

FIGURE 5.7 Transfer characteristics for high drain biases with varying temperature from 233 to 313 K. The channel length is 20 μm, and the drain bias is −30 V (a) and 30 V (b). The symbols represent the data points, and the solid line represents the theoretical fit. [*Source:* Smits *et al.* [37]. Adapted with permission from American Physical Society.]

5.2 FUNDAMENTAL EXCITONIC PROCESSES AND MODELING OF OLETs

Oppositely charged carriers (e.g., holes and electrons) can recombine in an organic solid releasing energy, which can be emitted in the form of light. This process is called radiative recombination, and in case electrons and holes are injected from electrodes into an organic system, it is named organic electro-luminescence (EL).

The charge-recombination process can be defined as fusion of a positive (e.g., hole, h) and a negative (e.g., electron, e) charge carrier into an electrically neutral entity or, following its evolution, in an electronic excited states. The initial (IR)—or geminate (GR)—and volume-controlled (VR) recombination can be distinguished by considering the charge carrier's origin [42]. GR is the recombination process following the initial carrier separation from a locally excited state, forming a nearest-neighbor charge-transfer (CT) state. It typically occurs as a part of intrinsic photoconduction in organic solids due to generation of charges from light-excited molecular states. The probability of GR (P_{GR}) can be expressed by the primary (often assumed to be electric-field independent) quantum yield in carrier pairs for the absorbed photon, η_0, and the (e–h) pair dissociation probability (Ω): $P_{GR} = 1 - \eta_0 \Omega$. The effect of the electric field on the charge separation can be experimentally observed in generation-controlled photoconduction or electric field-induced quenching of luminescence, because the external electric field promotes exciton dissociation and decreases the number of emitting states [43].

If the oppositely charged carriers are generated independently far away from each other (e.g., injected from distant electrodes), the volume recombi-nation takes place, and the carriers are statistically independent of each other forming electrically neutral entity (e.g., molecular exciton) [23]. The recom-bination process is kinetically bimolecular in this case.

A natural intermediate state before localization of both carriers on one molecular site is a bound pair, which is composed by an electron and a hole located at a distance $r < r_c$, where r_c is the distance at which the Coulombic binding energy equals the thermal energy. Indeed, the critical radius r_c can be defined as

$$r_c = \frac{e^2}{4\pi\varepsilon\varepsilon_0 kT} \qquad (5.28)$$

with ε being the constant dielectric of the organic solid, ε_0 the vacuum per-mittivity, k the Boltzmann constant, and T the temperature.

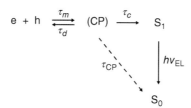

FIGURE 5.8   Three-step scheme of volume recombination, taking into account the formation of correlated e–h pair (CP) and a radiative decay of a localized singlet state (S_1).

The bound state can release its excess energy by direct transition (radiative or nonradiative) to the ground state (S_0) with a time constant τ_{CP}, by dissociation into a pair of free carriers (e, h) with a dissociation time constant τ_d or by the final recombination event (capture) with a characteristic capture time τ_c. In the latter step, excited molecular states (singlets S_1 and/or triplets T_1) are produced, leading to molecular fluorescence and/or phosphorescence (Figure 5.8).

Thus, two factors limit the radiative recombination of the charge carriers, the relaxation rate of the material, which above all determines how luminescent the material is, and the probability of charges with opposite signs to capture each other, which is defined by $P_c = (1 + \tau_c / \tau_d) - 1$.

There are two limiting cases for the volume recombination [44]: (i) the Thomson-like case, where the rate is limited by the phonon emission process, and (ii) the Langevin-like case, where the rate-limiting step is the diffusion of oppositely charged particles. Quantitatively, these two cases can be distinguished from each other by comparing the capture time τ_c with the carrier motion time τ_m, that is, the time to get the carriers within the capture radius. Thompson-like recombination happens when $\tau_c >> \tau_m$, and Langevin-like recombination when $\tau_c << \tau_m$. The different volume-recombination models correspond to the classical semiconductor approach if the mean free path for optical phonon emission, λ, is compared with the average distance $4r_c/3$ across a sphere of critical radius r_c [45]. However, a simple extrapolation of this approach to organics can be misleading because r_c (and even $r < r_c$) does not determine the capture cross section for the carriers.

Given the low carrier mobility in organic solids, the mean free path for elastic scattering λ (~10Å) is usually much lower than r_c (~150Å), thus suggesting that a Langevin-like model is more appropriate to describe the recombination process. A Langevin-like model is characterized by a field- and temperature-independent ratio between the recombination rate constant and the charge mobility [46]:

$$\frac{\gamma}{\mu_e + \mu_h} = \frac{q}{\varepsilon_0 \varepsilon_r} = \text{const} \tag{5.29}$$

where γ is the bimolecular (second-order) recombination rate constant, μ_e and μ_h are the electron and hole mobilities, respectively, and ε_r is the dielectric constant. Equation 5.32 is derived from the Smoluchowski expression relating the bimolecular recombination rate constant to the sum particle diffusion coefficient and their interaction radius once it is assumed the validity of Einstein's relation. Indeed, it highlights that the long-time limit charge recombination is a process controlled by diffusion. For organic molecular solids typically $\varepsilon_r = 4$, $\gamma/\mu = 4.5 \times 10^{-7}$ V cm, which, considering carrier mobilities of 10^2 cm^2/V s in the case of some aromatic crystals at low temperature [47] and of 10^{-10} cm^2/V s in the case of very low-mobility polymeric films, [48] span γ between 4.5×10^{-5} cm^3/s and 4.5×10^{-17} cm^3/s.

As we have discussed in the previous paragraph, organics can be generally considered as disordered solids where (i) the carrier motion is only partially diffusion controlled, and (ii) carrier hopping across a manifold of molecular sites statistically distributed in energy and space has to be taken into account [49]. The long-time balance between recombination and drift of carriers, expressed by the γ/μ ratio, has been analyzed using a Monte Carlo simulation technique. Since it has been demonstrated that γ/μ is to be considered independent from disorder [50], the Langevin formalism would be expected to obey recombination even in disordered molecular systems. However, the time evolution of γ is of crucial importance if the ultimate recombination event proceeds on the scale comparable with that of carrier pair dissociation. The recombination rate constant becomes then *capture* rather than *diffusion* controlled, so that Thomson-like model would be more adequate than Langevin-type formalism for the description of the recombination process. Nonetheless, the bimolecular carrier recombination is explained adequately in organic materials by simple Langevin theory, given that they can be considered as narrowband conductors in solid state and with charge-carrier mobility value of less than 1 cm^2/V s [24].

Whether this formulation for the bimolecular recombination rate constant may be used for describing the recombination rate process in OLET is still a topic of ongoing research.

While most optical devices such as OLEDs are essentially one-dimensional in nature, a light-emitting transistor requires the nonlinear and coupled exciton equations to be solved in two dimensions, in addition to Poisson's and drift–diffusion equations, given the intrinsic planar geometry.

Moreover, continuity equations for singlet and triplet excitons are to be implemented to account for the balance between generation and annihilation of excitons. In the exciton continuity equations, the Langevin term R is intended as a source for excitons, while radiative (r) and nonradiative (nr) decay, intersystem crossing (ISC), triplet–triplet (TT) annihilation, as well as singlet–singlet (SS), singlet–triplet (ST), singlet–polaron (SP), and triplet–polaron (TP) quenching mechanisms are included.

The singlet and triplet continuity equations form a set of equations that can be solved in addition to Poisson and charge-carrier continuity equations. Generally, the exciton dissociation is not considered in these equations [51,52].

$$\varepsilon_r \nabla^2 \psi = -q(p-n)$$

$$-\frac{1}{q}\nabla J_p - R = 0, \qquad -\frac{1}{q}\nabla J_n - R = 0$$

$$\chi R + \frac{\chi}{1+\chi}K_{TT}T^2 - \frac{S}{\tau_{rs}} - K_{ISC}S - K_{ST}ST - K_{SP}S(n+p) - \frac{K_{SS}}{2}S^2$$

$$- K_{nrs}S - \nabla(D_s \nabla S) = 0$$

$$(1-\chi)R + K_{ISC}S - \frac{T}{\tau_{rT}} - K_{TP}T(n+p) - K_{TT}T^2 - K_{nrT}T - \nabla(D_T\nabla T) = 0$$

$$(5.30)$$

where ψ is the electric potential, n and p are the electron and hole concentrations, S and T are the singlet and triplet concentrations, D_S and D_T are the diffusion constants of singlets and triplets, χ is the fraction of singlets (assumed to be ¼), k are the net rate constants, τ_r is the radiative lifetime, q is the elementary charge, and ε_r is the dielectric constant of the organic semiconductor.

This set of equations is implemented by Verlaak et al. [51] (together with a Schottky model for charge injection at the contacts) for determining the balance among the exciton decay, quenching, and generation mechanisms within a tetracene-based OLET. Indeed, the aim of the investigation was the comprehension of the main quenching mechanism, which may prevent the achievement of lasing in crystalline tetracene.

It was found that the singlet–triplet annihilation is the process that mainly affects the emission efficiency in single-layer OLET, given that the triplet exciton concentration is more than two orders of magnitude larger than that of singlet excitons.

On the other hand, Baldo et al. [52] demonstrated that in a device architecture based on single crystal, singlet–singlet annihilation is the main source of EL quenching. Indeed, the high mobility achieved in ordered molecular crystals is directly correlated to the lowering of the polaron density (at a given current density) and to the increase of the rate of the triplet losses.

These findings indicate that the singlet–triplet annihilation is reduced in molecular crystals. Band transport in crystalline materials further reduces singlet–triplet and singlet–polaron losses at low temperature where the field and thermally assisted hopping in amorphous materials freezes out triplet diffusion and retards charge transport as the temperature is decreased. On the contrary, losses in amorphous organic materials are mainly caused by high densities of triplets and polarons.

When the organic material that composes the active region of the OLET is polycrystalline, the scenario of the optoelectronic processes becomes more and more complex. Indeed, it is demonstrated that exciton–exciton and exciton–carrier quenching rates are strong functions of the exciton and carrier densities in OLETs so that both the maximum attainable exciton density and the internal and external quantum efficiencies of OLETs strongly depend on the width of the recombination zone (see Section 5.3).

Various approaches have been proposed in the literature for modeling the specific optoelectronic characteristics of OLETs in a given device configurations. A model is reported by Smith and Ruden [53] that allows for not only different electron and hole mobility but also charge trapping at the interface between the organic semiconductor and the gate dielectric, which is considered a fundamental issue in charge transport in organics [54].

In the ambipolar transport regime in OFETs, the charge carrier density is small in the crossover region between the electron- and hole-dominated transport because of the strong carrier recombination. Indeed, carrier trapping at the organic semiconductor/dielectric interface can be particularly relevant in low carrier-density device structures when charge trapping is not saturated. Carrier trapping can be described by treating the carrier mobilities as charge-carrier density dependent with low mobility value at low density. Since the charge-carrier density depends on the local voltage drop across the gate insulator, charge mobility is consequently modeled as voltage dependent.

This model, which is a generalization of device models developed for long-channel carbon nanotube light emitters [55,56], is implemented for describing polymeric light-emitting transistors. Interestingly, the functional voltage dependence of the mobility has not to be defined analytically and is not to be extended to arbitrarily low values.

If the current is assumed to be drift-dominated in the long-channel approximation frame, the recombination position x_0 in the channel can be defined as

$$x_0 = \frac{F(x_0)}{F(x_0) + \frac{\mu_h}{\mu_e} G(x_0)} L \qquad (5.31)$$

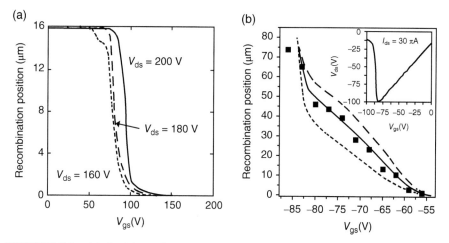

FIGURE 5.9   (a) Position of maximum recombination x_0 as a function of applied gate bias V_{gs} for three different source–drain voltage bias values according to the model developed by Smith *et al.* (b) Position of the recombination zone against applied gate bias V_{gs} according to the macroscopic model developed by Zaumseil *et al.* [58]. Measured values as squares, predicted values as: dotted line for a hole/electron mobility ratio of 0.2, solid line for a hole/electron mobility ratio of 0.4, dashed line for a hole/mobility ratio of 0.6. [*Source*: Zaumseil *et al.* [58], Figure 1, p 70. Adapted with permission of Nature Publishing Group.]

where F and G are integral functions, which depend on the local electrical potentials in the channel [53]. In order to verify the validity of the model, experimental device parameters such as gate-to-channel capacitance, channel length, drain–source bias, and high-carrier-density mobility values are needed. Particularly, good agreement with the experimentally observed results is found for the polymeric OLET implementing polyphenylene vinylene as active material and calcium (gold) as electron (hole)-injecting electrode in a top contact/bottom gate configuration [57].

Figure 5.9a shows the dependence of the recombination position x_0 on V_{gs} for several V_{ds} values.

At small gate bias, recombination occurs close to the source electrode (source for electrons), since relatively few electrons are injected and most of them have very low mobility due to the presence of traps. Only when the gate bias reaches a level that ensures that almost all traps are saturated, will the injected electrons be able to propagate into the channel and the recombination locus can move away from the source. At high gate bias, the low density of injected holes implies that they populate trap states near the drain (source for the holes).

It is interesting to note that formally similar results can be obtained for the determination of the recombination position and of the ambipolar OLET electrical characteristics by using less-refined macroscopic models [58,59].

Once it is assumed that the device channel in ambipolar conditions can be considered as the series of two n-type and p-type transistors in saturation, Zaumseil *et al.* [58] simply modeled the position of the light emission under given bias condition in a single-layer polymeric OLET. The fact that the experimental electrical characteristics followed the typical quadratic behavior for the saturation current was considered sufficient for implementing the well-established macroscopic theory of unipolar operating field-effect transistors [60].

The position of the recombination zone can be expressed as

$$x_0 = \frac{L\left(V_{gs} - V_{t,e}\right)^2}{\left(V_{gs} - V_{t,e}\right)^2 + \frac{\mu_h}{\mu_e}\left(V_{ds} - \left(V_{gs} - V_{t,h}\right)\right)^2} \tag{5.32}$$

Knowing the dimensions of the device, the approximate $V_{t,e}$ and $V_{t,h}$, and the ratio of hole to electron mobility (μ_h/μ_e), it is possible to model the position of the light emission in constant current mode.

The equation confirms the dependence of the recombination zone position on the applied voltages and on the ratio of the hole- and electron-mobility values. Figure 5.9b shows the good accordance between the position of recombination zone for different values of V_g as predicted by this model and that experimentally determined for the single-layer poly [2-methoxy-5-(3',7'-dimethyloctyloxy)-1,4-phenylenevinyle (MDMO-PPV) OLET [58].

Other simple analytical models can be implemented for determining the basic electrical characteristics as well as the potential and charge-carrier distribution in the channel of ambipolar field-effect transistors. Schmechel *et al.* [59] derived a model that may be considered as a generalization of the classical Shockley theory [60] for ambipolar transistors. Indeed, this model was developed by considering the device active channel as an equivalent circuit consisting of the resistors–capacitors combination reported in Figure 5.10a.

It consists in determining the channel resistance $R'dx$ for a differential segment dx, neglecting contact resistances and diffusion phenomena. The resistance varies according to the amount of charge carriers locally available for the transport:

$$R'(x)dx = \frac{dx}{Wq\left[\mu_n n(x) + \mu_p p(x)\right]} \tag{5.33}$$

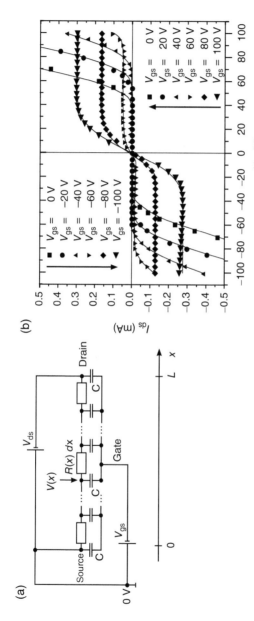

FIGURE 5.10 (a) Equivalent resistor–capacitor circuit for a thin-film field-effect transistor. (b) Output characteristics of an ambi-polar operating pentacene field-effect transistor. The symbols are the measured results, while the solid lines are model fits. [*Source:* Schmechel *et al.* [59]. Adapted with permission of AIP Publishing LLC.]

Here, $n(x)$ and $p(x)$ are the number of electrons and holes per unit area at the location x, while μ_n and μ_p define their respective mobility. W represents the channel width and q the elementary charge.

The net charge density in the channel can be approximated by the density of electrons and holes, independent of the net charge sign, since the concentration of electron–hole pairs is low due to high recombination probability.

The model was fitted to experimental electrical characteristics obtained from single-layer ambipolar pentacene OFET with gold and calcium top electrodes for guaranteeing Ohmic-like injection for both holes and electrons, respectively.

Figure 5.10b presents the output characteristics of an ambipolar pentacene field-effect transistor (symbols), as well as the fitted model (solid lines).

The output characteristics reported in the figure are quite symmetric for positive and negative drain voltages. It can therefore be assumed that the injection and transport properties are similar for both electrons and holes. The derived transistor model is able to describe the observed transistor characteristics, but the model parameters obtained from the fitting of the experimental data yield a strong threshold voltage shift for an increasing gate voltage. Since the voltage threshold parameter was not defined analytically, only a phenomenological explanation of the voltage-dependent behavior was given within the model.

The models for ambipolar OFETs that we have presented so far are based either on suitable modification of Pao–Sah [61] description for unipolar FETs or on the gradual channel approximation [37,53]. Thus, some physical and photophysical processes, which are fundamental for the complete description of ambipolar OLETs (nonohmic charge injections, charge recombination, etc.), are introduced ad hoc in the aforementioned models. Indeed, a certain degree of inaccuracy is generally tolerated in the model outputs, provided that a more exhaustive insight into the analytical relationships among the involved parameters is gained.

Another strategy to describe the ambipolar organic (light-emitting) transistors is to use state-of-the-art two-dimensional (2D) numerical simulations. The 2D simulation is required, due to the intrinsic planar architecture of the OFET.

Simulations have been carried out using the drift–diffusion model implementing two-dimensional device simulation programs such as ISE-TCAD [62] and Synopsys Taurus [51,63]. The application of this kind of device simulation software to organic materials was typically demonstrated in diode-like device structures [64–66] while several investigations on electrical characteristics of OFETs were performed as well [67–69].

The program solves simultaneously the Poisson equation for the electrica potential ϕ and the continuity equations for the hole (p) and electron (n) densities. In the nondegenerate limit, the latter are connected with the hole and electron quasi-Fermi potentials $\phi_{F,p}$ and $\phi_{F,n}$, respectively, by $p = n_i \exp[e(\phi_{F,p} - \phi)/k_B T]$ and $n = n_i \exp[e(\phi_{F,n} - \phi)/k_B T]$. The intrinsic density $n_i = (N_V N_C)^{1/2} \exp[-E_g/k_B T]$ is connected with the energy gap E_g and the effective densities of states N_V and N_C. For a molecular material, the molecular or monomer density has to be used rather than the effective DOS [66].

Ambipolar organic (light-emitting) transistors have been investigated in top contact/bottom gate device architectures with single-layer [62,70] and vertical heterojunction active regions [63,71].

Indeed, 2D numerical simulations of a single-layer OFET based on pentacene led to a useful understanding of the operation principles of such devices, clarifying the decisive influence of the nature of the contacts and the sensitivity to variations in the material parameters. Furthermore, the possibilities and restrictions of parameter extraction using simple analytical expressions were demonstrated.

In general, the nondegenerate limit of an extended Gaussian disorder DOS is adopted for explaining the self-consistent operation of pentacene-based OFETs, which is qualitatively assumed for a range of organic device structures reported in the literature [29]. Moreover, simulation studies are reported [72] in which broader Gaussian DOS caused slight deviations of the dependence of the accumulation charge per unit area on the gate voltage with respect to the nondegenerate approximation.

In addition, direct recombination has to be taken into account, because describing experimental data with models without recombination—if at all possible—inevitably leads to inappropriate values for the other model parameters. Generally, the bimolecular recombination rate in organic materials with low mobility is described within the Langevin approximation [24,73].

Finally, given the usual long-channel transistor approximation, the dependence of the mobilities on the electric field can be neglected and is not considered in the simulations.

In the simulations, one can vary material parameters in order to clarify systematically their influence on the current voltage characteristics or to achieve agreement with experimental data: that is, introducing a series resistance for taking into account the possible contact resistance, interface charge for describing the possible trapping at the organic–dielectric interface, or variable energy barriers at the electrodes for discriminating between Ohmic and Schottky-like injection processes.

Thus, Paasch *et al.* [62,71] clearly demonstrated ambipolar current–voltage characteristics comparable to the experimental ones measured in organic field-effect devices based on both single-layer and bilayer vertical hetero-structures. Also, optoelectronic characteristics of trilayer-heterostructure-based OLET were efficiently described by similar 2D simulations taking into account charge recombination and light formation [63].

Indeed, the difference between the organic field-effect devices based on a single layer and on a vertical heterojunction is also demonstrated. With specific regard to bilayer heterostructure obtained by superimposing p-transport and n-transport organic layers as device active region [74], the simulation was able to clarify the influence on the device characteristics of different electron and hole mobilities in both layers and of the band offset at the interface of the two layers. Particularly, it was possible to predict that the formation of a dominating electron channel at the interface between the n- and p-transport organic layers is to occur only when electron mobility in the n-transport layer is orders of magnitude larger than the hole mobility in the first layer.

However, the qualitative difference of the simulated current–voltage char-acteristics with respect to the experimental ones was attributed to the fact that the control of the second channel by the gate voltage is strongly reduced by the screening effect of the first charge channel at the interface with the dielectric. Indeed, several processes are neglected in the definition of the physical frame of the simulation, such as rechargeable interface states, trap distributions, or a nonconstant mobility.

Thus, it remains to be clarified whether these effects cause the modeling–experimental discrepancy or more realistic material parameters would be enough to improve the outcome of the simulation.

5.3 EXCITONIC RECOMBINATION AND EMITTING AREA IN OLETs

The investigation of the width of the recombination zone has so far received little attention. As we have already discussed, in theoretical works, the bimo-lecular recombination rate is taken to be infinite so that the recombination width is zero [37,59,70].

If a large recombination rate is imposed to the system so that electrons and holes cannot pass each other without recombining, the device equations can be solved analytically. The recombination takes place at a single position, namely where the effective gate potential is zero and the electrical field diverges. The assumption that electrons and holes move independently along the channel, with the consequent model representation of the ambipolar OLET as parallel unipolar *n*-type and *p*-type transistors in saturation, is considered of little physical meaning. Smits *et al.* [37] recognized the impossibility of determining the recombination rate from the experimental current–voltage characteristics within the model they proposed.

It is therefore necessary to understand why the Langevin expression is deficient in determining the recombination rate and to provide an alternative. The differences between theoretical and experimental results in OLETs may be originated by the near two-dimensional transport in these devices. Greenham and Bobbert [75] showed, by using a variety of techniques, that in the limit of two-dimensional transport, the rate constant becomes weakly dependent upon charge density, but did not discuss the impact of these findings on the Langevin expression.

Interestingly, in OLETs and multilayer OLEDs, it is likely that carriers will be confined within a few nanometers at an interface so that the rate constant may not be well described by either the three-dimensional limit used by Langevin or the two-dimensional limit considered by Greenham and Bobbert [75]. For devices comprising electron- and hole-accepting active materials, there seems to be a broad consensus from experiments that the rate constant is smaller than that predicted by the Langevin expression. However, the mechanism for this modification is not commonly agreed upon and has variously been attributed to the spatial separation of carriers [76], trapping via energetic disorder [77], isolated deep states [78], or the slower carrier's transit to the heterointerface [79]. While these mechanisms are all physically plausible, there is no comparative study that allows building consensus on their practical effects.

In order to define at least the vertical dimension of the recombination zone from the interface with the dielectric, Smith and Ruden [53] implemented an exponential decay function with increasing distance from the insulator/semiconductor interface for describing the spatial profile for the electrons and holes.

The recombination rate constant can be described as

$$\gamma = \frac{e\left(\mu_n + \mu_p\right)}{2\varepsilon_0\varepsilon_r\gamma_0} = \frac{e\left(\mu_n + \mu_p\right)G}{2\varepsilon_0} \tag{5.34}$$

The screening of the Coulomb attraction between electrons and holes is described by a dielectric constant ε. Since the confinement of the charge carriers in the direction perpendicular to the semiconductor/insulator interface is not well known, $G = 1/(\varepsilon_r y_0)$ is considered as a parameter in the expression of the vertical recombination width.

In regions where both electrons and holes populate the channel, recombination takes place over a length scale governed by

$$W = \sqrt{\frac{e(\mu_p + \mu_n)}{\gamma C_i}} = \sqrt{\frac{2d}{\varepsilon_r}} \sqrt{(1/G)} \qquad (5.35)$$

where as usual ε and t are the dielectric constant and the thickness of the gate insulator, respectively. In the previous simplified model [70], it was considered the limit $G \rightarrow \infty$, so that $W \rightarrow 0$.

In general, G is a functional parameter that it is supposed to take into account the decay in the electron and hole concentration profile with increasing distance from the dielectric surface.

A theoretical approach based on the traditional Langevin mechanism was proposed by Kemerink *et al.* [80]. It assumed that the carrier densities in the recombination zone are determined by the recombination process only. Having introduced a recombination rate using the Langevin bimolecular recombination rate constant and the calculated carrier densities, they defined the recombination width W as the value where the recombination rate reaches $1/e$ of its maximum value and find

$$W = \sqrt{4.34 d \delta} \qquad (5.36)$$

where d is the thickness of the gate insulator and δ the thickness of the accumulation layer.

Both the approaches yielded comparable results, since the recombination width depends on dielectric-related parameters and on the accumulation layer thickness. They both predict that the maximum extension of the recombination zone in single-layer OLET is around 500 nm. Nonetheless, experimental values for the recombination zone width do not confirm the predictions made by theory.

The in-plane architecture in OLETs allows direct probing of the different electronic and optical processes on the basis of the light generation. Moreover, deep investigation of the several functional interfaces present in the OLET structure is required to correlate film morphology and microstructure with fundamental processes such as charge injection from metal electrode into the organic layer and the current flow at the dielectric/organic interface [81].

Different characterization techniques are valuable tools for shedding new light into these fundamental issues. Local microscopy techniques such as scanning near-field optical microscopy, confocal microscopy, and scanning Kelvin probe microscopy are able to resolve the emission and potential profiles within the channel, thus mapping the organic/dielectric interface. Multiphoton time-resolved spectroscopy, such as electric-field-induced second-harmonic generation [82], can also help unraveling charge- and electric-field-induced quenching phenomena inside working OLETs.

Zaumseil *et al.* [83,84] reported for single-layer F8BT-based OLETs an emission zone width of 2 μm, which was measured by optical microscopy.

Laser scanning confocal microscopy (LSCM) is a powerful imaging technique to obtain highly spatially resolved imaging of the shape and dimensions of the emission area *W*. Indeed, the highest possible resolution is achieved by collecting light at or near the diffraction limit ($\lambda/2$) and refocusing it in the confocal plane by means of a pinhole. Therefore, only light originating from the focal point is collected by the detector, whereas all light emanating from out-of-focus regions is rejected. Unlike conventional microscopy, the collected light forms a point source at the confocal plane and provides no spatial resolution. Spatial resolution is obtained by scanning the confocal spot over the sample, effectively creating an image point by point. Combining this scanning technique with incident laser excitation provides a high-resolution, high-sensitivity optical imaging system [85].

Swensen *et al.* [86] interfaced an encapsulated polymer-based single-layer OLET with a laser scanning microscope. Prior to device operation, the channel region was imaged with the confocal microscope. The channel region was defined by a region of high photoluminescence intensity about 16 μm long with regions of no photoluminescence on both sides. The bright PL region corresponds to the luminescent semiconducting polymer in the channel. The dark regions on both sides of the channel are the silver and calcium electrodes; the metal electrodes block the laser from exciting the underlying luminescent polymer.

Once the confocal microscope objective was positioned over the channel, the laser illumination was turned off, and the OLET was operated in the dark. Any EL within the channel was detected by the confocal microscope as it was scanned back and forth across the channel of the OLET. Under these experimental conditions, the highly resolved light signal is the EL emitted within the device channel.

A bias condition was set and held constant for a given period of time while the confocal microscope imaged the emission. Once a sufficiently long time had elapsed such that a good image was obtained, the bias was turned off, and the next bias was then set, applied, and held constant (Figure 5.11).

It is interesting to note the tilt of the collected emission lines at longer collection times (Figure 5.11a) as a sort of slow creep of the emission line toward the calcium electrode on the right side of the channel. This phenomenon is probably due to bias-induced threshold voltage shift for electron transport. An analysis of the spatial intensity profile shows that the full width at half maximum of the emission zone is consistently 2 μm as it moves across the channel.

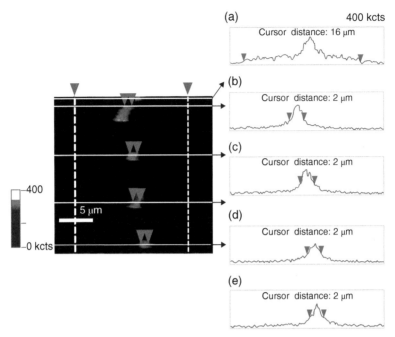

FIGURE 5.11 Confocal microscopy images of the emission zone collected as it moved across the channel region of the OLET. Cross section plots of emission intensity versus lateral position in each scan are shown on the right. Cursors and the extended dotted lines depict the location of the electrode edges that define the channel region. Silver is the left electrode and calcium is the right electrode. (a) $V_d=115\,$V, $V_s=-115\,$V, and $V_g=0\,$V. (b) $V_d=115\,$V, $V_s=-111\,$V, and $V_g=0\,$V. (c) $V_d=115\,$V, $V_s=-109\,$V, and $V_g=0\,$V. (d) $V_d=115\,$V, $V_s=-105\,$V, and $V_g=0\,$V. [*Source*: Swensen *et al.* [86]. Adapted with permission of AIP Publishing LLC.]

A study using scanning Kelvin probe microscopy (SKPM) gives some support to the theoretical predictions because it indicates a smaller recombination zone width with values down to $0.5\,\mu$m [87]. SKPM measurements rely, however, on heavy post-treatment of the data in order to reduce instrumental artifacts and, therefore, should be taken into consideration with care. The magnitude of the disagreement between the measured width of the emission zone and the calculated width of the recombination zone cannot be due to exciton diffusion only, since the diffusion length is usually in the range of tens of nanometers. To reproduce values of W in the micrometer range by using the previously explained models, the bimolecular recombination rate constant γ has to be significantly decreased below the Langevin value [80]. This reduction, in turn, leads to a smaller exciton density and consequently to a wider emission zone. For a phenomenological description, one can modify γ by introducing a simple numerical prefactor α in the Langevin bimolecular recombination rate constant:

$$\gamma = \alpha \frac{q}{\varepsilon_0}\left(\mu_e + \mu_h\right) \tag{5.37}$$

Here, $\alpha = 1$ stands for the normal Langevin recombination rate. Reducing γ by a factor α leads to a reduction of the recombination zone width by a factor $\sqrt{\alpha}$. Thus, to reproduce the measured width of $2\,\mu m$, α has to be of the order of 0.01–0.001.

As we mentioned before, one possible reason for such a reduction in α is the two-dimensional (2D) nature of the charge transport within the accumulation layer. Simulations [84] demonstrated that restricting the transport to 2D slows down recombination and reduces α to 0.05. Moreover, the same study revealed a strong correlation between the microstructure of the semiconducting film and the width of the recombination zone. Indeed, charge transport in the direction of F8BT polymer chains considerably broadens the emission zone width (up to $10–12\,\mu m$) with respect to the case of transport taking place perpendicularly to the chain direction (Figure 5.12).

In addition, mobility values measured along or perpendicular to the chain direction comprise a clear anisotropy of 11 for holes and 7 for electrons. These findings show that the reduction of the recombination rate is strongly influenced by the mobility anisotropy. However, the Langevin recombination process implemented for the modeling of OLETs assumes only isotropic mobility. By taking into account the mobility anisotropy and 2D carrier motion, the recombination rate can be reduced by a factor of 100. However, this value should be reduced by an additional factor of 1000 to corroborate the experimental data. It is clear that a full comprehension of the microscopic mechanisms at the basis of the light-generation process is still missing, even in the case of the simplest single-layer/single-material OLET, and further theoretical and experimental investigations are needed.

The optical images of the emission zone sweeping across the channel in a bottom gate/bottom contact polymer-based OLET during a transfer mode scan is reported in Figure 5.13 [58]. The OLET fabricated by Zaumseil *et al.* presented a solution-processed conjugated polymer, poly(2-methoxy-5-(3, 7-dimethyloctoxy)-*p*-phenylene-vinylene) (OC1C10-PPV or MDMO-PPV) as active material, silicon dioxide with a buffer layer of cross-linked benzo-cyclobutene derivative (BCB) as dielectric, and gold and calcium electrodes as source and drain contacts, respectively.

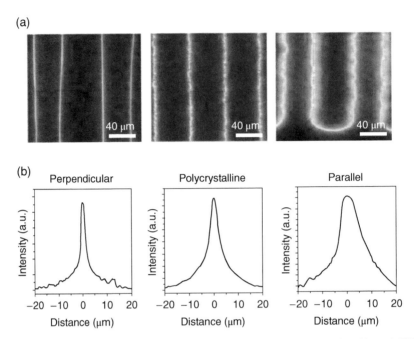

FIGURE 5.12 Emission zone for perpendicularly and parallelly aligned F8BT OLET. Optical microphotographs (a) and intensity profiles (b) of the emission zone. [*Source*: Zaumseil *et al.* [84]. Adapted with permission from Wiley.]

Below a certain $|V_{ds}|$, no ambipolar transport or light emission is observed as the channel is dominated by hole transport and no electrons are injected. When $|V_{ds}|$ increases, the relative bias between drain and gate exceeds the threshold voltage for formation of an electron accumulation layer. A narrow line of light appears alongside the edge of the electron-injecting calcium electrode. Owing to the presence of trapped holes at the interface, the threshold voltage for electron accumulation is in fact negative; that is, in Figure 5.13a, light can already be observed when V_{ds} is still less negative than V_{gs}. With increasing V_{ds}, the emission zone moves through the channel toward the gold electrode as the electron accumulation region extends further into the channel. With increasing voltage, and thus increasing current, the light becomes brighter. At higher $|V_{ds}|$, the light is located close to the gold electrode and continues to increase in intensity. When $|V_{ds}|$ is reduced again, the light moves back to the calcium electrode. The intensity profiles do not show any strong asymmetry or variations in peak width with intensity or position.

It has been demonstrated that the light emission region features depend not only on the structural properties of the active layer material but also on the overall optoelectronic device characteristics in OLETs. Indeed, EL signal in

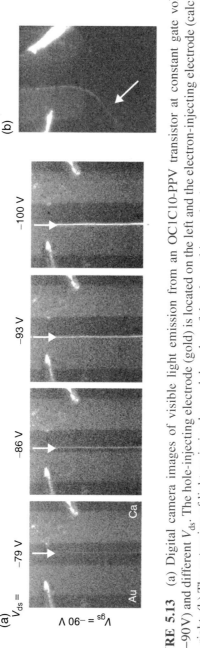

FIGURE 5.13 (a) Digital camera images of visible light emission from an OC1C10-PPV transistor at constant gate voltage ($V_{gs} = -90\,V$) and different V_{ds}. The hole-injecting electrode (gold) is located on the left and the electron-injecting electrode (calcium) on the right. (b) The extension of light emission beyond the edge of the channel in a pulsed measurement switching between $V_{ds} = 0$ and $-100\,V$ and $V_{gs} = -90\,V$. [*Source:* Zaumseil *et al.* [58], Figure 4, p. 73. Adapted with permission of Nature Publishing Group.]

different transistor regimes has been used as a probe to characterize the properties of the different functional interface, such as metal/organic interface. Schidleja *et al.* [88] showed that the effect of the charge-carrier injection on the performance of ambipolar light-emitting OFETs can be investigated by implementing devices with completely different behavior concerning the bias-dependent light emission. Indeed, Figure 5.14 reports the transfer characteristics (at constant $V_{ds} = 100\,V$) together with the position of the light emission in the channel for three different single-layer OLETs as a function of the sweeping gate voltage. Particularly, the three device architectures are as follows: top gate/bottom contact F8BT-based polymeric OLET implementing PMMA as dielectric and gold as both source and drain electrodes (Figure 5.14a); bottom gate/top contact OLET with ditetracene small molecule as active material, poly(vinyl cyclohexane) as dielectric, and gold as both source electrodes (Figure 5.14b); and the same bottom gate/top contact ditetracene-based device with the optimized metal contacts for electron and hole injection (i.e., calcium contact for electron injection and gold contact for hole injection) (Figure 5.14c).

The F8BT-based OLET shows the highest light intensity in the ambipolar regime, while in the ditetracene devices, the highest light output appears in the unipolar n-type and p-type regimes, respectively. A comparison of the two ditetracene-based OLETs, whose setups only differ in the contact material used for the electron injection, indicates that the origin of the different device behavior is due to the contact formation at the source and drain electrodes.

By detecting the light emission from OLETs in the unipolar regimes, the contact properties of the injecting electrode can be characterized. In the case of ditetracene-based devices, the formation of ohmic contact at the metal/organic interfaces is expected, given the suitable alignment between the work function of Au and the HOMO of ditetracene and the work function of Ca and the LUMO of ditetracene. Indeed, this energetic condition is the cause of the formation of charge reservoirs in the proximity of both source and drain contacts resulting in light emission far away from the electrodes. As it can be seen from Figure 5.14b and c, the intensity of EL signal can be directly correlated to the position of light formation given that metal electrodes act as exciton quenchers. In contrast, the fading light emission in the unipolar regimes of F8BT transistor indicates no or little thermal charge transfer at the metal/semiconductor interface, in agreement with Mott–Schottley model.

For multilayer heterostructure OLETs, even more complex models have to be implemented for describing the charge-transport and carrier-recombination processes. In fact, the emission zone width in trilayer heterostructure OLETs [89] is very broad and can reach 70–100 μm [63].

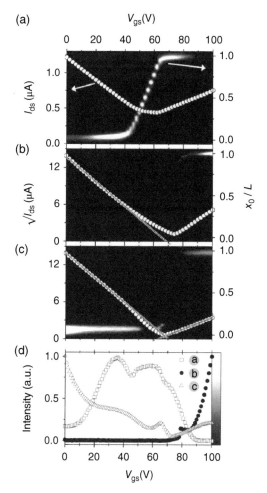

FIGURE 5.14   Transfer characteristics (open circles) of different ambipolar OLETs at a constant drain voltage of $V_{ds} = 100$ V and the position of the recombination zone x_0 (arrows for indicating) are reported. The pixel brightness in the $x_0(V_{gs})$ characteristics in (a)–(c) equals the emitted light intensity shown in (d). For coding, the respective shades are displayed on the right side of (d). For better comparability, in (d), the emitted intensities from the three devices are compared. The device structures are as follows: (a) Top gate, bottom contact: Au (source, drain)/F8BT/ PMMA/Au (gate). (b) Bottom gate, top contact: Si-p++(gate)/SiO$_2$/PVCH/ditetracene/ Au (source, drain). (c) Bottom gate, top contact: Si-p++(gate)/SiO$_2$/PVCH/ ditetracene/Ca (source)/Au (drain). [*Source*: Schidleja *et al.* [88]. Adapted with permission of AIP Publishing LLC.]

(a) (b) (c)

(d) (e) (f)

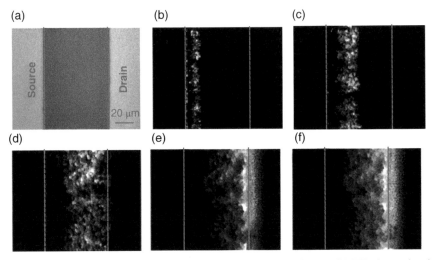

FIGURE 5.15 Images of the light-emitting area in the trilayer OLET channel collected through the glass substrate. A transmission optical microphotograph of the unbiased channel is reported for reference (a). V_{gs} is varied (−20 V (b), −40 V (c), −60 (d), −80 V (e), and −100 V (f)), while V_{ds} is kept constant at −100 V. The intensity scale of the different optical images is normalized to avoid image saturation. [*Source*: Toffanin *et al.* [63]. Adapted with permission from Wiley.]

Indeed, it was recently demonstrated that in the suitably optimized trilayer OLET architecture, not only the position but also the extension of the emission area is controlled by the applied voltage, achieving the illumination of the entire transistor channel. By increasing the field-effect charge-transport properties of the top conducting layer and the electron and hole mobility balance (which are strongly correlated to the channel illumination characteristics), it was possible to control the width of the emission area in OLETs.

In Figure 5.15, the optical microphotographs of the device channel when the trilayer OLET is biased at constant V_{ds} and different V_{gs} values are reported.

Figure 5.15a shows the transmission optical image of the device channel area for reference. The EL signal is collected through the transparent ITO/ glass substrate. Applying $V_{gs} = -20$ V and $V_{ds} = -100$ V (Figure 5.15b), therefore working in the electron-injection regime, a 10 μm wide light-emitting region is formed in the channel starting from the edge of the source electrode. With increasing $|V_{gs}|$, the emission stripe shifts toward the drain electrode, becoming broader and brighter. In the ambipolar regime ($V_{gs} = -60$ V, $V_{ds} = -100$ V, Figure 5.15d), the emission area covers almost completely the 70 μm channel, reaching the drain electrode and extending for few microns underneath. The intensity maximum is located in the middle of the channel with an almost symmetric profile. Shifting from ambipolar to complete unipolar p-transport regime, the EL maximum is peaked in the proximity of the drain electrode and the emitting area extends underneath the electrode for 20 µm. The emission in the center of the channel is progressively reduced as $|V_{gs}|$ increases above 60 V. However, the emission zone exceeds 40 µm also in the purely unipolar p-transport condition ($V_{gs} = -100\,\text{V}$, $V_{ds} = -100\,\text{V}$, Figure 5.15f).

In order to gather information about the specific charge-recombination process at the basis of the unprecedented broadening of the emission area in three-layer OLETs, the lateral and vertical exciton distributions in the OLET channel were modeled for the different bias conditions. Indeed, the authors reported that the extension of the emission area is mainly due to the vertical electrostatic interaction between flowing hole and electron currents with consequent distributed charge injection and hopping into the recombination layer where the charge recombination is mainly localized. Moreover, in the trilayer architecture, the homogeneously distributed emission area within the channel in ambipolar bias conditions is strongly correlated to the balance of hole and electron current density in both the bottom and top transport layers.

This picture was confirmed by the 2D numerical simulation of the trilayer heterojunction transistor. The two-dimensional simulations are performed by means of a commercial device simulator adapted for including the Langevin recombination rate in the continuity equations for holes and electrons (Figure 5.16).

By decrease in the absolute value of the applied source–gate voltage, the excitons in the channel are progressively confined at the interface between the recombination and the n-transport layer where the maximum of the emitted power is invariably localized. The optical emission broadens underneath the drain electrode for $V_{gs} \leq -80\,\text{V}$ (Figure 5.16a and b). This behavior is in agreement with the experimental observations and demonstrates that the electrostatic interaction among accumulated charge distributions of opposite signs is on the basis of the exciton formation process. The experimentally and theoretically demonstrated gate-tunable broadening of the emission zone is a distinguishing nontrivial feature of the trilayer device architecture, which is capable of avoiding the in-plane interpenetration of the electron and hole flowing currents, differently from single-layer-based device.

It is interesting to note that the use of simple pn junction model cannot explain all the different experimental determinations of the emission (recombination) zone width even in single-layer ambipolar transistors model systems. Particularly, studies performed on thin-film transistors have not yet revealed much of the recombination physics in ambipolar OLETs indicating

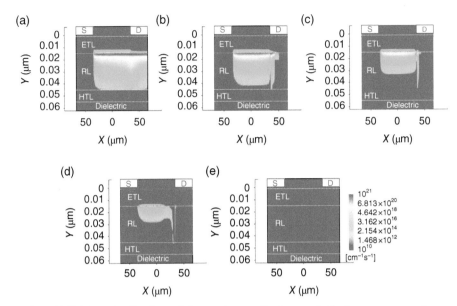

FIGURE 5.16 Distribution of the exciton radiative annihilation rate per volume unity within a transversal section along the channel length of a trilayer OLET, when the device is biased at $V_{gs}=-100\,V$ (a), $V_{gs}=-80\,V$ (b), $V_{gs}=-60\,V$ (c), $V_{gs}=-40\,V$ (d), and $V_{gs}=-20\,V$ (e), respectively, while keeping $V_{ds}=-100\,V$. Color legend is also reported. [*Source*: Toffanin *et al.* [63]. Adapted with permission from Wiley.]

that there are still hidden parameters and unknown physics in determining analytically the recombination zone width.

Once more, the single-crystal OLETs can provide model systems to shed light on these open issues. Birsi *et al.* [90] measured the recombination width (W) of ambipolar OLETs by varying the organic single crystals: tetraphenyl-pyrene (TPPy), 4,4′-diphenyl-vinylene-anthracene (DPVA), BP3T, and rubrene. The ambipolar OLETs were fabricated as a bottom-gate top-contact device, where their hole and electron injections have been optimized by the use of Au and Ca electrodes, respectively. The emission/recombination zone was probed by using an optical microscope mounted with CCD cameras with different sensitivities.

Figure 5.17 shows the optical micrographs of the emission zone in four different ambipolar OLET devices with bis(biphenylyl)terthiophene BP3T, rubrene, DPVA, and TPPy single crystals.

All of the emission zones were clearly visible as line shapes. The full width at half maximum (FWHM) value of the light emission intensity profile collected across the transistor channel represents the emission zone

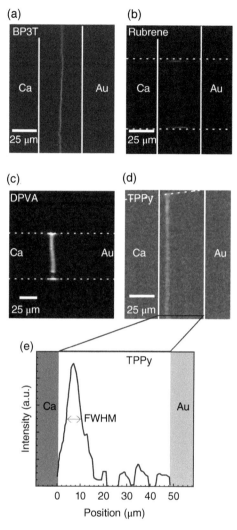

FIGURE 5.17   (a) Light emission from ambipolar OLETs using a single crystal of BP3T, (b) rubrene, (c) DPVA, and (d) TPPy operated in a V_{ds} bias sweep mode. (e) An intensity profile of light emission across the channel length of a TPPy single-crystal device for the recombination zone width determination. [*Source*: Bisri *et al.* [90]. Adapted with permission from Wiley.]

width (Figure 5.17e). Assuming that the recombination zone size is proportional to that of the emission zone W, values obtained from several different OLETs with four different materials were less than 10 μm wide and narrower than the emission zone size in tetracene single-crystal ambipolar OLET, which was reported by Takahashi ($W = 12$ μm) [91]. Indeed, some of

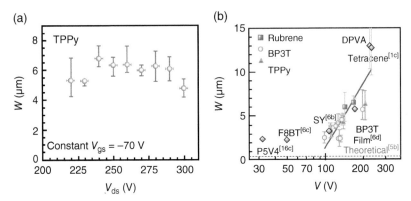

FIGURE 5.18   (a) The dependence of the recombination zone width on the applied source–drain bias in single-crystal OLET based on TPPy, in the range when the recombination zone is moving across the transistor channel. (b) The relationship between W and V, which indicates a logarithmic trend, is in good agreement with the pin junction model. [*Source*: Bisri *et al.* [90]. Adapted with permission from Wiley.]

the values were even comparable to the sizes obtained in ambipolar OLETs based on polymer and small-molecule thin films. However, the smallest W value obtained from single-crystal OLETs is more than 2 µm, which is still much larger than that of theoretical estimation from a pn junction modeling of the single-layer single-material OLET. After having excluded a possible overestimation of emission zone originated by optical dispersion and wave-guiding effect, the pn model was considered not suitable to describe the observed variety of W.

Indeed, it was demonstrated that W did not increase with increasing diffusion rate of the charge carriers at the interpenetrating boundary, as it is expected in pn diode junction. This can be easily evaluated from the invariance of the recombination zone width with respect to increasing source–drain bias at constant source–gate bias as reported in Figure 5.18.

Moreover, it was observed that W was not narrowing with increasing charge-carrier mobility as expected, according to the bimolecular recombination theory, which is normally assumed in pn junction model.

Particularly, the recombination zone width values in rubrene and tetracene [91] OLETs are wider than those of BP3T- and TPPy-based devices, whereas the electron and hole mobilities in rubrene and tetracene OLETs are comparable to those of BP3T and two to three orders of magnitude higher than that of TPPy. This discrepancy strongly indicates that W observed in the single-crystal ambipolar OLETs cannot be rationalized in terms of the intrinsic recombination process in the material.

Birsi *et al.* proposed a pin junction model for describing the origin of light formation within single-crystal-based devices, where an intrinsic (undoped) region is located within the hole- and electron-accumulation layers. It is well known that carrier accumulation only occurs in the channel where the local effective gate bias is zero. If there is a significant threshold voltage (as it is typically expected in single-crystal devices) for either hole or electron accumulation, there will be no charge accumulation in the proximity of the recombination zone where the effective voltage is lower than the threshold voltage. Thus, the ambipolar single-crystal OLET can be considered as *pin* junction device where the depletion zone width is almost gate-bias independent given that the charge-carrier diffusion only occurs at its boundaries.

The *pin* model is based on the drift–diffusion transport mechanism, where the relationship between the applied voltage V and the width of the undoped region W can be expressed by

$$V = \frac{2kT\mu_p\mu_n}{q(\mu_p + \mu_n)^2}\exp\left[\frac{W}{2L_a}\right], \text{ when } \frac{W}{L_a} > 2 \qquad (5.38)$$

L_a is the ambipolar diffusion length defined as $(D_a\tau_a)^{1/2}$, where D_a is the ambipolar diffusion constant and τ_a is the ambipolar lifetime [92].

Equation 5.41 can be easily rearranged considering that the recombination zone width is much larger than the typical charge-carrier diffusion length in organic materials and the mobility value is strongly dependent only on the density of trap states in single crystal.

Indeed, a logarithmic functional dependence of W from the applied bias V can be found, which can be considered in the undoped region as the sum of the electron and hole threshold voltage values, which is well corroborated by experimental findings.

Nonetheless, it is important to highlight that the pin model can be fruitfully applied only to single-crystal-based OLETs.

5.4 CONCLUSIONS

In this chapter, the charge-transport theories, excitonic processes, modeling of the optoelectronic characteristics, and emitting area features of OLETs were introduced and discussed. It is worth highlighting that, although the physical analysis is built on that of consolidated devices such as OLEDs and OFETs, many aspects of the OLET behavior remain to be investigated. Indeed, despite a first order level of overall agreement, a number of open issues emerge when the electrical and optoelectronic characteristics are analyzed at a deeper level. We anticipate that further fundamental investigations and understanding will be of crucial importance for the next development phase of OLETs and for the optimization of their photonic properties. The photonic properties of OLETs are the subject of Chapter 6.

REFERENCES

[1] H. Bassler, *Phys Status Solidi.* **1993**, *175*, 15.

[2] E. A. Silinsh and V. Càpek: *Organic Molecular Crystals: Interaction, Localization, and Transport phenomena* (AIP Press, New York, **1994**).

[3] M. C. J. M. Vissenberg and M. Matters, *Phys. Rev. B* **1998**, *57*, 12964.

[4] J. L. Bredas, D. Beljonne, J. Cornil, J. P. Calbert, Z. Shuai, and R. Silbey, *Synth. Metal* **2002**, *125*, 107.

[5] R. C. Haddon, X. Chi, M. E. Itkis, J. E. Anthony, D. L. Eaton, T. Siegrist, C. C. Mattheus, and T. T. M. Palstra, *J. Phys. Chem. B* **2002**, *106*, 8288.

[6] Y. C. Cheng, R. J. Silbey, D. A. Da Silva, J. P. Calbert, J. Cornil, and J. L. Bredas, *J. Chem. Phys.* **2003**, *118*, 3764.

[7] G. Horowitz, *J. Mater. Res.* **2004**, *19*, 7.

[8] S. F. Nelson, Y. Y. Lin, D. J. Gundlach, and T. N. Jackson, *Appl. Phys. Lett.* **1998**, *72*, 1854.

[9] G. Horowitz, M. E. Hajlaoui, and R. Hajlaoui, *J. Appl. Phys.* **2000**, *87*, 4456.

[10] G. Horowitz, R. Hajlaoui, R. Bourguiga, and M. Hajlaoui, *Synth. Metal* **1999**, *101*, 401.

[11] A. S. Dhoot, G. M. Wang, D. Moses, and A. Heeger, *J. Phys. Rev. Lett.* **2006**, *96*, 246403.

[12] A. S. Dhoot, *et al.*, *Proc. Natl. Acad. Sci. USA* **2006**, *103*, 11834.

[13] J. D. Yuen, *et al.*, *Nature Mater.* **2009**, *8*, 572.

[14] V. N. Prigodin and A. Epstein, *J. Phys. Rev. Lett.* **2007**, *98*, 259703.

[15] J. H. Worne, J. E. Anthony, and D. Natelson, *Appl. Phys. Lett.* **2010**, *96*, 053308.

[16] T. Sakanoue and H. Sirringhaus, *Nature Mater.* **2010**, *9*, 736.

[17] N. Karl, J. Marktanner, *Mol. Cryst. Liq. Cryst.* **2001**, *355*, 149.

[18] O. Ostroverkhova, D. G. Cooke, S. Shcherbyna, *et al.*, *Phys. Rev. B* **2005**, *71*(3), 035204.

[19] N. A. Minder, S. Ono, Z. Chen, *et al.*, *Adv. Mater.* **2012**, *24*, 503.

[20] N. Narl, J. Marktanner, R. Stehle, and W. Warta, *Synth. Metal* **1991**, *42*, 2473.

[21] S. Fratini and S. Ciuchi, *Phys. Rev. Lett.* **2009**, *103*, 266601.

[22] A. Troisi and G. Orlandi, *Phys. Rev. Lett.* **2006**, *96*, 086601.

[23] I. N. Hulea, S. Fratini, and H. Xie, *Nature Mater.* **2006**, *5*, 982.

[24] M. Pope and C. E. Swenberg, *Electronic Processes in Organic Crystals and Polymers* (Oxford University Press, New York, **1999**).

[25] C. X. Liang and M. D. Newton, *J. Chem. Phys.*, **2000**, *112*, 1541.

[26] N. Tessler and Y. Roichman, *Organic Field Effect Transistors*, Ed. J. C. deMello and J. J. M. Halls (John Wiley and Sons, **2003**).

[27] C. Tanase *et al.*, *Phys. Rev. Lett.* **2003**, *91*, 216601.

[28] H. Sirringhaus, N. Tessler, and R. H. Friend, *Synth. Met.* **1999**, *102* 857.

[29] S. Shaked, S. Tal, Y. Roichman, A. Razin, S. Xiao, Y. Eichen, and N. Tessler, *Adv. Mater.* **2003**, *15*, 913.

[30] Y. Roichmann, Y. Preezant, and N. Tessler, *Phys. Stat. Sol. A* **2004**, *201*, 1246.

[31] P. G. Le Comber and W. E. Spear, *Phys. Rev. Lett.* **1970**, *25*, 509.

[32] G. Horowitz, R. Hajlaoui, and P. Delannoy, *J. Phys. III France* **1995**, *5*, 355.

[33] A. R. Völkel, R. A. Street, and D. Knipp, *Phys. Rev. B* **2002**, *66*, 195336.

[34] A. Dieckmann, H. Bassler, and P. M. Borsenberger, *J. Chem. Phys.* **1993**, *99*, 8116.

[35] E. J. Meijer, C. Tanase, P. W. M. Blom, E. van Veenendaal, B. H. Huisman, D. M. de Leeuw, and T. M. Klapwijk, *Appl. Phys. Lett.* **2002**, *80*, 3838.

[36] C. Tanase, E. J. Meijer, P. Blom, and D. de Leeuw, *Org. Electron.* **2003**, *4*, 33.

[37] E. C. P. Smits, T. D. Anthopoulos, S. Setayesh, E. van Veenendaal, R. Coehoorn, P. W. M. Blom, B. de Boer, and D. M. de Leeuw, *Phys. Rev. B* **2006**, *73*, 205316.

[38] A. F. Stassen, *et al.*, *Appl. Phys. Lett.* **2004**, *85*, 3899.

[39] V. Ambegaokar, B. I. Halperin, and J. S. Langer, *Phys. Rev. B* **1971**, *4*, 2612.

[40] M. Sahimi, *Applications of Percolation Theory* (Taylor & Francis, London, **1994**).

[41] A. R. Brown, C. P. Jarrett, D. M. de Leeuw, and M. Matters, *Synth. Met.* **1997**, *88*, 37.

[42] J. Kalinowski, *J. Phys. D: Appl. Phys.* **1999**, *32*, R179.

[43] J. Kalinowski, W. Stampor, and P. Di Marco, *J. Electrochem. Soc.* **1996**, *143*, 315.

[44] J. Kalinowski, *Mol. Cryst. Liq. Cryst.* **2001**, *355*, 231.

[45] R. Morris and M. Silver, *J. Chem. Phys.* **1969**, *50*, 2969.

[46] P. Langevin, *Ann. Chim. Phys.* **1903**, *28*, 289.

[47] J. S. Brooks, *Chem. Soc. Rev.* **2010**, *39*, 2667.

[48] G. Juska, K. Arlauskas, J. Stuchlik, and R. Osterbacka, *J. Non-Cryst. Solids* **2006**, *352*, 1167.

[49] M. Kuik, G. J. A. H. Wetzelaer, H. T. Nicolai, N. I. Craciun, D. M. De Leeuw, and P. W. M. Blom, *Adv. Mater.* **2014**, *26*, 512.

[50] J. J. M. van der Holst, F. W. A. van Oost, R. Coehoorn, *et al.*, *Phys. Rev. B* **2009**, *80*, 235202.

[51] S. Verlaak, D. Cheyns, M. Debucquoy, V. Arkhipov, and P. Heremans, *Appl. Phys. Lett.* **2004**, *85*, 2405.

[52] M. A. Baldo, R. J. Holmes, and S. R. Forrest, *Phys. Rev. B* **2002**, *66*, 035321.

[53] D. L. Smith, P. P. Ruden, *J. Appl. Phys.* **2007**, *101*, 084503.

[54] L. L. Chua, J. Zaumseil, J. F. Chang, E. C. W. Ou, P. K. H. Ho, H. Sirringhaus, and R. H. Friend, *Nature* **2005**, *434*, 194.

[55] J. Tersoff, M. Freitag, J. C. Tsang, and P. Avouris, *Appl. Phys. Lett.* **2005**, *86*, 263108.

[56] J. Guo and M. A. Alam, *Appl. Phys. Lett.* **2005**, *86*, 023105.

[57] J. S. Swensen, C. Soci, and A. J. Heeger, *Appl. Phys. Lett.* **2005**, *87*, 253511.

[58] J. Zaumseil, R. H. Friend, and H. Sirringhaus, *Nature Mat.* **2006**, *69*, 5.

[59] R. Schmechel, M. Ahles, and H. von Seggern, *J. Appl. Phys.* **2005**, *98*, 084511.

[60] G. Horowitz, *Adv. Mater.* **1998**, *10*, 365.

[61] G. W. Neudeck and H. F. Bare, *IEEE Trans. on Electron Devices* **1987**, *34*, 344.

[62] G. Paasch, T. Lindner, C. Rost-Bietsch, S. Karg, W. Riess, and S. Scheinert, *J. Appl. Phys.* **2005**, *98*, 084505.

[63] S. Toffanin, R. Capelli, W. Koopman, G. Generali, S. Cavallini, A. Stefani, D. Saguatti, G. Ruani, and M. Muccini, *Laser Photonic Rev.* **2013**, *7*, 1011.

[64] S. Scheinert, G. Paasch, P. H. Nguyen, S. Berleb, and W. Brütting, *Proc ESSDERC'00* (Editions Frontiers, Paris, **2000**), 444.

[65] P. H. Nguyen, S. Scheinert, S. Berleb, W. Brütting, and G. Paasch, *Org. Electron.* **2001**, *2*, 105.

[66] A. Nesterov, G. Paasch, S. Scheinert, and T. Lindner, *Synth. Met.* **2002**, *130*, 165.

[67] T. Lindner, G. Paasch, and S. Scheinert, *J. Mater. Res.* **2004**, *19*, 2014.

[68] S. Scheinert and G. Paasch, *Phys. Status Solidi A* **2004**, *201*, 1263.

[69] S. Scheinert, G. Paasch, M. Schrödner, H. K. Roth, S. Sensfuß, and T. Doll, *J. Appl. Phys.* **2002**, *92*, 330.

[70] D. L. Smith and P. P. Ruden, *Appl. Phys. Lett.* **2006**, *89*, 233519.

[71] T. Linder, G. Paasch, and S. Scheinert *J. Appl. Phys.* **2007**, *101*, 014502.

[72] G. Paasch and S. Scheinert, *J. Appl. Phys.* **2007**, *101*, 024514.

[73] P. W. M. Blom and M. J. M. de Jong, *Philips J. Res.* **1998**, *51*, 479.

[74] C. Rost, D. J. Gundlach, S. Karg, and W. Riess, *J. Appl. Phys.* **2004**, *95*, 5782.

[75] N. C. Greenham and P. A. Bobbert, *Phys. Rev. B* **2003**, *68*, 245301.

[76] I. Balberg, R. Naidis, M. K. Lee, J. Shinar, and L. F. Fonseca, *Appl. Phys. Lett.* **2001**, *79*, 197.

[77] G. J. Adriaenssens and V. I. Arkhipov, *Solid State Commun.* **1997**, *103*, 541.

[78] J. Nelson, S. A. Choulis, and J. R. Durrant, *Thin Solid Films* **2004**, *451*, 508.

[79] L. J. A. Koster, V. D. Mihailetchi, and P. W. M. Blom, *Appl. Phys. Lett.* **2006**, *88*, 052104.

[80] M. Kemerink, D. S. H. Charrier, E. C. P. Smits, S. G. J. Mathijssen, D. M. de Leeuw, and R. A. J. Janssen, *Appl. Phys. Lett.* **2008**, *93*, 033312.

[81] M. A. Loi, E. Da Como, F. Dinelli, M. Murgia, R. Zambomi, F. Biscarini, and M. Muccini, *Nature Mat.* **2005**, *4*, 81.

[82] T. Manaka, E. Lim, R. Tamura, and M. Iwamoto, *Nature Photon.* **2007**, *1*, 581.

[83] J. Zaumseil, C. R. McNeil, M. Bird, D. L. Smith, P. P. Ruden, M. Roberts, M. J. McKiernan, R. H. Friend, and H. Sirringhaus, *J. Appl. Phys.* **2008**, *103*, 064517.

[84] J. Zaumseil, C. Groves, J. M. Winfield, N. C. Greenham, and H. Sirringhaus, *Adv. Func. Mater.* **2008**, *18*, 3630.

[85] K. D. Weston, P. J. Carson, H. Metiu, and S. K. Buratto, *J. Chem. Phys.* **1998**, *109*, 7474.

[86] J. S. Swensen, J. Yuen, D. Gargas, S. K. Buratto, and A. J. Heeger, *J. Appl. Phys.* **2007**, *102*, 013103.

[87] E. Smits, S. Mathijssen, M. Colle, A. Mank, P. Bobbert, P. Blom, B. de Boer, and D. de Leeuw, *Phys. Rev. B* **2007**, *76*, 125202.

[88] M. Schidleja, C. Melzer, M. Roth, T. Schwalm, C. Gawrisch, M. Rehahn, and H. von Seggern *Appl. Phys. Lett.* **2009**, *95*, 113303.

[89] R. Capelli, S. Toffanin, G. Generali, H. Usta, A. Facchetti, and M. Muccini, *Nature Mater.* **2010**, *9*, 496.

[90] S. Z. Bisri, T. Takenobu, K. Sawabe, S. Tsuda, Y. Yomogida, T. Yamao, S. Hotta, C. Adachi, and Y. Iwasa *Adv. Mater.* **2011**, *23*, 2753.

[91] T. Takahashi, T. Takenobu, J. Takeya, and Y. Iwasa, *Adv. Funct. Mater.* **2007**, *17*, 1623.

[92] S. M. Sze, *Physics of Semiconductor Devices*, 2nd ed., (Wiley, New York, NY, USA, **1981**).

6

PHOTONIC PROPERTIES OF OLETs

In combination with the monolithic integration of multiple electronic and optoelectronic functions, which are guaranteed by the device architecture, the photonic properties of organic light-emitting transistors (OLETs) are essential for any practical applications. The efficiency of the device structure in terms of photons generated per injected charge and the maximum optical power at a given electrical bias are key parameters to evaluate the OLET devices from the luminosity and light efficacy point of view. In this chapter, these photonic characteristics of OLETs are highlighted, together with the light outcoupling and emission directionality properties. The novel opportunities offered by the planar device geometry and the lateral charge transport characteristics in the search for organic injection lasing are also discussed.

6.1 EXTERNAL QUANTUM EFFICIENCY

One of the most important figures of merit in terms of light generation for every light-emitting device is the external quantum efficiency (EQE). It gives a measure of the degree of conversion of flowing current within the device into emitted light. The EQE can simply be defined as the ratio of the photon

Organic Light-Emitting Transistors: Towards the Next Generation Display Technology,
First Edition. Michele Muccini and Stefano Toffanin.
© 2016 John Wiley & Sons, Inc. Published 2016 by John Wiley & Sons, Inc.

flux per unit area measured outside the device with respect to the carrier flux per unit area ζ injected into the device:

$$\eta_{ext} = \zeta\gamma^{\Phi}{}_{PL}\chi^{\sigma}{}_{out} \tag{6.1}$$

where γ is the ratio of the number of exciton formation events within the device with respect to the number of charges flowing in the external circuit, Φ_{PL} is the luminescence quantum yield of the exciton formation layer, χ is the spin multiplicity of the radiatively recombining excitons, and σ_{out} is the light outcoupling efficiency from the device into the open space.

In general, the process of electron–hole capture (which is at the basis of the exciton formation) is crucial to optoelectronic performances in organic light-emitting devices. Indeed, to obtain an efficient capture in the in-plane micrometer-wide recombination volume typical of OLETs, it is necessary to guarantee a sufficiently high charge density so that charges of opposite signs can encounter each other within the collisional capture radius [1].

Indeed, in single-layer ambipolar OLETs, we can expect that all carriers recombine since charges cannot travel over the distance of several micrometers in the accumulation layer of the opposite charge without recombining [2].

As long as light is emitted from the internal region of the channel, at a distance from the source and drain edges that prevents contact interference, EQE of ambipolar light-emitting transistors is expected to be constant and should only depend on the singlet/triplet ratio, radiative yield of formed excitons, and light outcoupling efficiency irrespective of, for example, the ratio of hole and electron mobility or voltage conditions.

Typically, the value of γ is considered to be 1, even though it is strictly correlated to the effective balance between hole and electron mobility (see the following text).

It is known from organic light-emitting diode (OLED) technology that in vertical multilayer-based devices, the electron–hole recombination rate is maximized by the charge confinement within the heterojunction with consequent buildup in charge density [1].

Indeed, it has been demonstrated that this assumption holds true also in three-layer OLET devices for the charges entering the recombination layer [3]. Experiments and simulations show, in the case of three-layer OLETs, that most of the flowing charge carriers percolate in the middle layer, with consequent enhancement of the EQE up to 5%.

The fraction of radiative species is largely dominated by the singlet–triplet ratio in the emissive layer. For example, in fluorescent materials, only singlet states are emissive; thus, all triplet states are lost for EL. Out of quantum-statistical reasons, charge recombination yields 25% singlet excitons only [4], which can be furthermore reduced by *static quenching*. This kind of quenching includes typically charge trapping at the semiconductor/dielectric interface and electrode-induced quenching.

The luminescence quantum yield is the fraction of generated excitons that actually do relax radiatively. The luminescence quantum yield of the material itself, which also depends on intermolecular nonradiative decay channels, is further reduced by the interaction with quenching species.

The so-called *dynamic quenching* includes exciton scattering, such as scattering of excitons with other excitons, free charge carriers (polarons), or defects and chemical impurities, as well as exciton dissociation into holes and electrons.

In principle, ambipolar OLETs should reduce, or even avoid, exciton–polaron scattering and quenching at the electrodes; both are known to be severe limitations for the efficiency of light-emitting diodes [5–7]. Yet, the influence of luminescence quenching in OLETs is disputed.

As we described in Chapter 5, an early numerical simulation [8] named singlet–triplet and singlet–polaron scattering as primary sources of singlet quenching in OLETs devices. This study did not include exciton dissociation due to the field effect in the recombination region, in consideration of the device geometry and of the field-effect charge transport. Indeed, the electric field intensity in the channel of a typical OLET presenting a 500-nm-thick dielectric layer remains below 10^6 V/cm for a gate bias up to 100 V. Such a value is usually considered too low to generate sizable exciton dissociation in a single-layer and single-material organic device [9]. Comparatively, it is worth noting that the dominating source of exciton quenching in vertically stacked devices such as OLEDs is found to be the field-enhanced exciton dissociation [10].

Another relevant issue to be considered is the effect of the interaction between carriers and excitons. In this process, the excitation energy is transferred via dipole–dipole interaction (Förster transfer) from the excited neutral molecule to a charged molecule and subsequently dissipates nonradiatively. In vertically stacked devices, the charge density is typically about 10^{15} cm^{-3}, which can be correlated to an average charge–charge distance of about 100 nm. Considering that the exciton diffusion length is in the range of tens of nanometers, exciton–charge interaction is negligible in devices with vertical stack geometry [11,12]. On the contrary, the sheet charge density can reach values up to 10^{14} cm^{-2} [13] in OFETs, which corresponds to a distance of 1 nm among charges, thus well within the diffusion length of excitons. However, theoretical models predict a very low effective charge-carrier density inside the recombination zone. Thus, the nonradiative decay connected with high current densities (e.g., exciton–polaron annihilation) is expected to be of minor importance. In addition, terahertz radiation measurements on nonemitting polymer-based field-effect transistors demonstrated a quenching of the radiation-induced excitons limited to 20% [14].

The current density for polymer OLETs, for example, reaches values of $50\,A/cm^2$ without a decrease of quantum efficiency [15]. Single-crystal devices allow for even higher densities up to $4\,kA/cm^2$ [16]. These density values are considerably higher than those achieved in conventional OLEDs, which are in the order of $1-10\,A/cm^2$ at their point of maximum efficiency.

For the sake of completeness, we mention that calculating the actual current density in a (single-layer) light-emitting transistor is not straightforward. It is not known how long holes and electrons remain confined to the interface within the recombination zone compared to their respective accumulation regions (within about $1-2\,nm$ of the interface [17]) or how distant they diffuse from the interface into the bulk, and, if so, how this depends on voltage conditions. Nevertheless, a lower and an upper limit for the current density in light-emitting transistors can be estimated by taking into account the current flow through an area that is defined by the channel width and the height of the emission zone. For a normalized source–drain current of $1\,\mu A/cm^2$, the corresponding upper limit is $10\,A/cm^2$ (assuming that all charges are confined to within $1\,nm$ at the active interface), while the lower limit of current density is calculated to be $0.2-0.14\,A/cm^2$ using the typical thickness of the polymeric or small-molecule active layer ($50-70\,nm$). However, this calculation does not yet take into account the width of the emission zone in OLETs ($2-4\,\mu m$), which indicates that charge recombination is less spatially confined in an OLET than in an OLED [18,19].

An interesting investigation about the emission characteristics and external quantum efficiencies as a function of the applied voltage, current density, and ratio of the hole to electron mobility in ambipolar polymer light-emitting field-effect transistors was provided by Zaumseil and coworkers [6]. The fabricated devices are bottom contact/top gate, and the active layers are the green-emitting poly(9,9-di-n-octylfluorene-alt-benzothiadiazole) (F8BT) with balanced electron and hole mobilities and the red-emitting polymer poly((9,9-dioctylfluorene)-2,7-diyl-alt-[4,7-bis(3-hexylthien-5-yl)-2,1,3-benzothiadiazole]-2′,2″-diyl (F8TBT) with strongly unbalanced hole and electron mobilities. Figure 6.1 shows the current–voltage and corresponding light output characteristics of F8BT and F8TBT OLETs together with their EQEs as a function of gate and source–drain voltages. The EQE of F8BT OLET reaches a plateau at around 0.55% centered around the source–drain current minimum for $V_{ds}=-100\,V$. However, at $V_{ds}=-80\,V$, the maximum efficiency is lower by almost 20%. The EQE, thus, seems to increase with increasing V_{ds}. Furthermore, for $V_{ds}=-80$ and $-90\,V$, a dip efficiency coincident

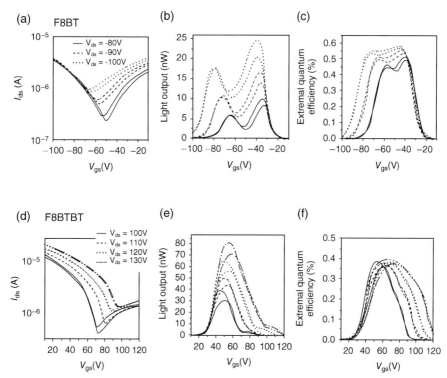

FIGURE 6.1 (a) Source–drain current, (b) light output, and (c) external quantum efficiency versus gate voltage (forward and reverse) of a light-emitting F8BT (green emission) transistor with $L=20\,\mu m$, $W/L=500$, $C_i=5.9\,nF/cm^2$, $\mu_e=5\times10^{-4}\,cm^2/V\,s$, $\mu_h=6\times10^{-4}\,cm^2/V\,s$. (d) Source–drain current, (e) light output, and (f) external quantum efficiency versus gate voltage (forward and reverse) of a light-emitting F8TBT (red emission) transistor with $L=20\,\mu m$, $W/L=1000$, $C_i=6.8\,nF/cm^2$, $\mu_e=3\times10^{-5}\,cm^2/V\,s$, $\mu_h=6\times10^{-4}\,cm^2/V\,s$. [*Source*: Zaumseil *et al.* [6]. Adapted with permission of AIP Publishing LLC.]

with the current minimum is evident when the emission zone is expected to be in the middle of the channel.

This pronounced dip of efficiency for the F8BT OLETs at lower V_{ds} together with the lowered efficiency at lower drain currents in the case of F8TBT OLETs suggests that the observed changes of EQE might be caused by different current densities rather than by the source–drain voltage or lateral field.

The calculated EQEs, in agreement with the measured data, display less structured profile than the light output versus gate voltage curves. Particularly, the strong asymmetric shape of the light output is not found in the EQEs of the

F8TBT device because the effect of the lower electron mobility is largely canceled out in the EQE

By means of the analytical model for organic ambipolar OLET developed by Smith and Ruden [19], the simulated EQE curves of these devices displays a flat plateau with increasing V_{gs} differently from what is observed experimentally. This discrepancy can be explained because the implemented model is limited to a single (Langevin) recombination mechanism, without taking into account other competing mechanisms, especially the excitonic processes with carrier density dependency [20].

In order to deconvolute the influences of V_{ds} and source–drain current on EQE in OLETs, constant current measurements are to be performed on both types of polymer transistors.

From the constant current measurements, it is possible to extract the light output and the EQE for a range of currents encompassing more than two orders of magnitude for both the investigated polymers (Figure 6.2). It is important to normalize the source–drain current against the channel width in order to be able to compare devices with different W/L ratios and perform a measurement of the current density. As it can be seen, the EQE of both types of OLETs initially increases strongly with increasing source–drain current and then starts to saturate approaching a constant efficiency at source–drain currents exceeding 1.5 µA/cm. In the case of the mobility-balanced polymer, the efficiency saturation value is $(0.8 \pm 0.1)\%$.

By performing simulation of the device optoelectronic based on Lu and Sturm model [21] (see Chapter 5) and suitable outcoupling efficiency calculation, it was found that, in the case of a device with a bottom contact and top

(a) 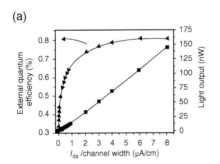 (b)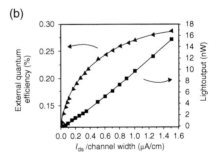

FIGURE 6.2 Light output (squares) and external quantum efficiency (triangles) versus channel-width-normalized source–drain current of (a) F8BT ($W = 1$ cm) and (b) F8TBT ($W = 2$ cm) transistors. Data points were extracted from plateau part of light output versus V_{gs} of constant current measurements. [*Source*: Zaumseil *et al.* [6]. Adapted with permission of AIP Publishing LLC.]

gate configuration with light extraction through the 15-nm-thick gold gate electrode, the maximum EQE achievable is 0.75%. Zaumseil *et al.* confirmed that the experimentally measured EQE of 0.8% is the evidence of very efficient OLET operation.

The increase of the quantum efficiency with current density up to a saturation value could be indicative of a trap-assisted nonradiative decay at the semiconductor–dielectric interface. Indeed, if a charge is trapped at the interface, or at a defect in the bulk of the emitting polymer, recombination with an opposite charge is likely to be nonradiative assuming that a charged trap is also a luminescence quenching site. As the source–drain current increases, the number of defects at the interface or in the bulk and, thus, the fraction of nonradiative recombination events become less significant compared to the number of radiative recombination events between mobile charge carriers, which results in increased EQE.

Experimental evidences show that EQE is dependent on current density in polymer-based single-layer OLETs: at low current densities, the EQE increases strongly and then approaches a saturation value, which is consistent with complete charge recombination. The maximum achievable EQE in single-layer OLET with a given device architecture is limited by the fundamental properties of the active material. Consistently, further investigation revealed that an improvement in EQE is achieved by optimizing the device architecture and electroluminescence collection strategy. Gwinner *et al.* [22] optimized the usual bottom contact/top gate OLETs based on the semiconducting polymer F8BT, which yield EQE values of about 8% and luminance efficiencies >28 cd/A. Particularly, the strategy consisted in (i) enhancing electron injection by functionalizing the electrode with ZnO, (ii) optimizing the PMMA dielectric thickness, (iii) using low-loss silver reflecting mirror as gate electrode, and finally (iv) extracting light through the glass substrate. Indeed, the main enhancement of EQE is due to the absence of high-refractive-index transparent electrode (i.e., ITO) in the OLET structure, which causes severe in-plane waveguiding of the emitted light (σ_{out} in Eq. 6.1). By collecting photoluminescence (PL) quenching profiles within the channel of a working single-layer OLET, the charge density in the region where EL occurs was found to be so low that the luminescence efficiency was unaffected by the presence of charges. Indeed, it was inferred experimentally that this condition was verified only in the center of the recombination zone, while at the edges of the recombination zone, charge-induced PL quenching is present. This consequence of the unique electrostatic features of ambipolar OLETs constitutes an important advantage over OLEDs. Indeed, the extreme spatial localization of charge carriers in an OLET could be more favorable for an effective spatial separation between the exciton population and the charge carriers. The availability of a third electrode to balance electron and hole currents and therefore to further reduce exciton–polaron quenching is the other obvious advantage of OLETs.

Nonetheless, under intense charge-injection and -transport conditions, which are necessary to guarantee high brightness, EQE roll-off with flowing current is expected, given that regions of carrier accumulation and exciton generation largely coalesce in single-layer devices.

For their OLET, Capelli *et al.* [3] replaced the single-layer-based organic semiconductor by a trilayer structure consisting of an emission layer sandwiched between an electron- and hole-transporting layer on a polymeric poly(methyl methacrylate) insulator (Figure 3.14). The authors attribute the resulting large EQE to a favorable situation within the emission zone, largely closing essential loss channels for electroluminescence. As the recombination zone is far away from any metal contact and from the insulator, metal-induced quenching and quenching at trapped charges at the insulator surface are hindered. Indeed, the separation of the light emission from the charge-carrier accumulation zones separates regions of large carrier densities from regions of large exciton densities, reducing exciton–polaron quenching.

The trilayer OLET devices reach 5% EQE and are >100 times more efficient than the equivalent OLED (Figure 6.3), >2 times more efficient than the optimized OLED with the same emitting layer.

From experimental and simulation evidences (Chapter 5), it was demonstrated that the excitons are confined between the recombination and the n-transport layers, thus highlighting the complete decoupling between the light-formation and the charge-transport zones. Thus, avoiding the in-plane interpenetration of the hole and electron flowing currents, the trilayer OLET is the ideal candidate for realizing organic light-emitting devices that will outperform current devices in terms of luminance and/or excitation density and enable new applications, such as high-brightness displays or electrical-injection lasers.

One of the strong points of the trilayer approach is that each layer, which is devoted to a specific function in the device (charge injection and transport, radiative exciton recombination), can be optimized independently. Clearly, matching the overall device characteristics with the functional properties of the single materials composing the active region of the OLET is a great challenge that requires a deep investigation of the morphological, optical, and electrical features of the system.

On the other hand, engineering a unique organic semiconductor capable of performing well-balanced ambipolarity and highly efficient light emission, as it is mandatory in the realization of single-layer OLET, is a task even more challenging from the synthesis and device fabrication standpoint. Since

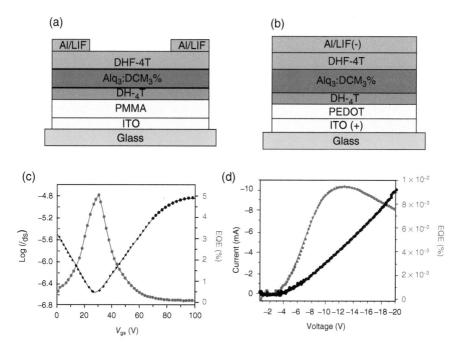

FIGURE 6.3 (a–c) EQE as a function of the applied gate voltage for the optimized trilayer heterostructure OLET configuration reported by Capelli *et al.* [3]. (b–d) Device structure and optoelectronic characteristics of the OLED corresponding to the trilayer OLET reported in (a). [*Source*: Capelli *et al.* [3], Figure 6, p. 501. Adapted with permission of Nature Publishing Group.]

this challenging goal is a highly rewarding one, significant effort has been spent in this direction, which has led to important progress in designing ambipolar polymer semiconductors with balanced electron and hole field-effect mobilities exceeding $1\,cm^2/V\,s$ [23].

Recently, it was demonstrated that small-molecule organic semiconductors can be implemented for realizing a single-layer molecular OLET that compares favorably with state-of-art single-layer ambipolar OLETs based on a polymeric active layer [24]. Indeed, it was demonstrated that the insertion of a thieno(bis)imide (TBI) moiety as end group into linear thiophene oligomers promotes ambipolar charge transport and electroluminescence in the resulting semiconductors. The data of Melucci *et al.* showed that substitution with one TBI end moiety can be exploited for (i) tuning the highest occupied molecular orbital (HOMO) energy values of oligothiophenes while maintaining unchanged the lowest unoccupied molecular orbital (LUMO) energy, (ii) localizing the LUMO distribution on the oligomer periphery, (iii) mastering the charge transport (ambipolar with predefined major p- or n-type behavior) by a proper TBI opposite end substitution, and (iv) enabling thin-film electro-luminescence in combination with ambipolar charge transport.

The EQE is not only related to the implemented organic semiconductor, being small-molecule or polymer, and thus to the specific molecular structure, but also to the supramolecular packing.

Particularly, the use of single crystals as the active layer generates high field-effect mobilities and high current densities that further brighten the light emission. In fact, the luminescence efficiency does not show roll-off up to several kiloamperes per square centimeter (assuming 1 molecular layer (1 ML) as the accumulation layer thickness) in tetracene and rubrene single-crystal OLETs [16]. Such high efficiency preservation is not shared by OLEDs, and single-crystal OLETs are thus considered preferable candidates for high-intensity light-emitting devices. It has been suggested that metal electrode effects, singlet–polaron annihilations (SPAs), singlet–singlet annihilations (SSAs), and/or singlet–triplet annihilations (STAs) may be possible origins of the efficiency roll-off [5,16]. Although the high current densities and lack of roll-off effects in tetracene and rubrene single-crystal OLETs are key evidence for clarifying the dominant mechanism, the luminescence efficiency of these light-emitting transistors is very low, and they are not best suited for studying optical properties.

Recently, it was demonstrated that single crystals of 5″-bis (biphenylyl)-2,2′:5′,2″-terthiophene (BP3T) in OLETs show high field-effect mobilities for both holes and electrons with current densities greater than $4\,kA/cm^2$ (assuming a 3 nm thick (1 ML) accumulation layer) [25,26]. It was then attempted to further increase the maximum current density to investigate the luminescence properties at higher current densities. A maximum current density of $33\,kA/cm^2$ was obtained, which is approximately one order of magnitude higher than the previous maximum current density achieved in single-crystal OLETs [27]. Moreover, a linear increase in the luminance was observed up to a current density of $1\,kA/cm^2$, while further current injection demonstrated the roll-off characteristics of BP3T single-crystal OLETs. This efficiency preservation is three orders of magnitude greater than that of conventional OLEDs at high current densities [5] (Figure 6.4).

There are many possible causes for this impressive difference in optoelectronic performances between OLEDs and (single-crystal) OLETs. The first cause is the effect of light absorption by the metal electrodes, which is negligible in ambipolar OLETs and has already been demonstrated for tetracene and rubrene OLETs [16]. Another possible explanation points to the properties of the specific materials employed to fabricate the OLET devices. BP3T is exceptionally less susceptible to SPA, SSA, and STA, which may explain the reported results. However, this is unlikely to be the major cause

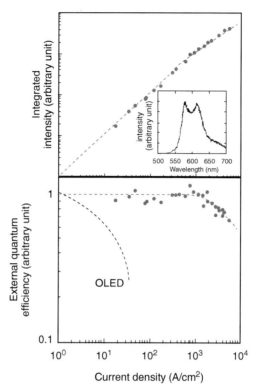

FIGURE 6.4 Current-density dependence of the integrated intensity of the luminescence spectra (top) and external quantum efficiency (bottom) in the ambipolar BP3T single-crystal OLETs. The inset shows a typical luminescence spectrum of the BP3T single-crystal OLETs during ambipolar transistor operation. The results observed for OLEDs are also plotted [5]. The dashed lines present a guide for the eye. [*Source*: Sawabe *et al.* [27]. Adapted with permission from Wiley.]

because the roll-off point occurs universally in the range of $100\,mA/cm^2$ for OLEDs made from several different materials [28,29].

The triple-digit gap between the EQE roll-off points of the OLEDs and OLETs (Figure 6.4) can be considered according to the authors as a clear consequence of the reduced SPA mechanism in OLETs.

Indeed, a difference of three orders of magnitude between the mobilities of the OLETs ($\approx 10^{-1}\,cm^2/V\,s$) and OLEDs ($10^{-3}$ to $10^{-4}\,cm^2/V\,s$) is present [30] so that the carrier concentration of the OLETs is far more dilute than that of the OLEDs at the same current density.

However, the density of the singlet and triplet excitons is proportional to the current density, and the situation is identical for both OLETs and OLEDs, which implies that STA [31,32] and SSA [33] are not the dominant quenching mechanisms. Therefore, the obtained difference between the OLETs and OLEDs might be interpreted as SPA.

The extremely high current densities and the roll-off characteristics are expected to extend the performance limits of OLETs and will open a realistic route toward bright organic lighting devices.

In any case, it is clear from this discussion that in order to maximize the optoelectronic performance of OLETs, recombination and photophysical processes within the devices have to be subjected to further intense investigations. In the following sections, we devote our attention to the brightness and light outcoupling characteristics of OLETs, which are strictly correlated to device architecture engineering.

6.2 BRIGHTNESS

To turn organic optoelectronic devices into a power-efficient light source, three key parameters must be addressed: the internal electroluminescence quantum efficiency must be close to 1 (high internal quantum efficiency), a high fraction of the internally created photons must escape to the forward hemisphere (high outcoupling efficiency), and the energy loss during electron–photon conversion should be small (low operating voltage).

The use of phosphors allows 100% internal quantum efficiency, because both the singlet and triplet states (generated at a ratio of 1:3 owing to their multiplicity) are directed to the emitting triplet state [34]. For power-efficient white OLEDs, an additional challenge is that high-energy phosphors demand host materials with even higher triplet energies to confine the excitation to the emitter [35]. Taking exciton binding energy and singlet–triplet splitting into account, the use of such host materials considerably increases the transport gap and therefore the operating voltage. For these reasons, blue fluorescent emitters are widely used to complete the residual phosphor-based emission spectrum [36–38]; this, however, either reduces the internal quantum efficiency or requires blue emitters with special properties [39]. Whenever OLEDs are built in a standard substrate-emitting architecture, the outcoupling efficiency is approximately 20%. The remaining 80% of the photons are trapped in organic and substrate modes in equal amounts [40]. Hence, the greatest potential for a substantial increase in EQE and power efficiency is to enhance the light outcoupling [41].

The brightness and light emission performance reported for OLETs in the literature are rather low. This is mainly because the organic materials used to date in OLETs possess either a low carrier mobility with high PL, such as the amorphous organic semiconductors used in OLEDs [15,42–44], or a high mobility with weak PL, such as in crystalline materials [45–47]. To achieve high performance, materials must (i) be capable of ambipolar carrier (hole/electron) injection and transport from their respective source–drain electrodes, (ii) have high PL efficiency in a thin film, and (iii) have relatively good carrier transport. Balancing these three factors in a single-layer OLET device structure is a challenging synthesis and device fabrication task.

Even though success with ambipolar OLETs [3,22] has been reported, high efficiency was mainly demonstrated only in the lowest current regime where the emitted light intensity is very weak. To obtain high energy efficiency, it is necessary to eliminate as much energy waste as possible. For example, one must minimize energy dissipation through gate leakage, or eliminate exciton quenching at contacts, or reduce contact resistance, and so on. If the operating voltage is high, which is typically the case for OLETs (>50 V) compared to OLEDs (<10 V), the only way to decrease the dissipation of energy is to reduce the operating current. However, in conventional OLETs, lower current means lower brightness, especially when the EQE is low.

Another important issue that must be properly taken into account when dealing with brightness is the experimental setup and protocol that must be implemented for obtaining reliable and effective measurements of luminance (or brightness in photometric units).

According to the handbooks of applied photometry, the luminance of a point source emitting in an elemental solid angle $d\Omega_S$ (Figure 6.5) is the radiant flux ($d\Phi$) per unit solid angle and per unit projected area perpendicular to the specific direction.

$$L = \frac{d^2\Phi}{d\Omega_s \, dA_s \, \cos\theta} \tag{6.2}$$

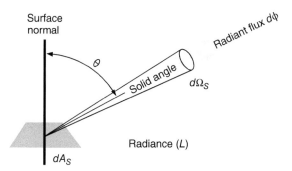

FIGURE 6.5 Schematics of the definition of luminance according to Equation 6.2.

where dA_S is the elemental area containing the given point, $d\Omega_S$ contains the given direction, and θ is the angle between the normal to the elemental area and the given direction.

This conventional definition of luminance holds true until the emitting source can be considered point-like. Real-world radiometric and photometric problems deal with extended sources (as in the case of organic optoelectronic devices); thus, it is usually supposed that extended sources can be treated as large collection of identical, uniformly distributed point sources.

The essential criterion for a source to be treated as a point-like source is that the product of the lateral dimensions is small compared to the square of the distance to the source. An ideal point-like source is the one that emits isotropically in the sense that the flux emitted is the same in all the directions. In the case of OLETs, this condition can easily be fulfilled given the micrometric (~10–100 µm) length of the active channel, despite the intrinsic anisotropy of the device planar geometry.

Moreover, for performing accurate and self-consistent luminous-intensity measurements of light sources, a photometric bench is necessary where calibration lamps or light sources and photodetector head are positioned and the source-to-detector distance is varied freely and measured accurately [48]. Indeed, the photometric bench is a complex optical setup often equipped with telescopes for aligning the light source along the optical axis and with several aperture screens for avoiding the situation that light may directly enter the photodetector from the space surrounding the light source. Finally, either standard lamps or standard photometers can be used as reference methods. Luminous-intensity standard lamps, calibrated in one horizontal direction, provide the luminous-intensity unit (candela). This kind of experimental protocol is implemented for measuring brightness in the case of extended superficial light sources such as OLED. Indeed, luminance meter and spectroradiometer, in particular, are well diffused in both academic and research and development laboratories.

However, direct measurements of brightness are not typically reported in the case of OLETs. Typically, the optoelectronic characterization of OLETs is held onto a probe station equipped with a photodetector for collecting light, which is housed in an atmosphere-controlled glovebox. Namdas *et al.* [49] reported a brightness value for an OLET, which was calculated by comparing the photocurrent in a photomultiplier collected from an extended-surface reference device of known brightness and light emission area [50], but without specifying the source–detector distance. Indeed, given that the emissive area of the reference device and of the OLET is different, the photocurrent has to be corrected for the effective OLET emissive area to obtain the brightness value. It is evident that by scaling for the typical small emitting areas of OLETs, very intense values of brightness are obtained.

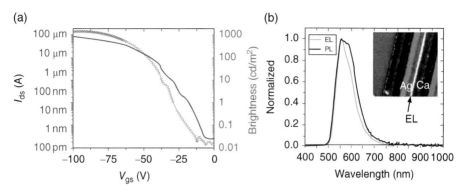

FIGURE 6.6 (a) Electrical (blue) and optical (red) transfer characteristics of the OLET. The gate voltage was scanned from 0 to −100 V while keeping the source–drain voltage at a fixed value of −100 V. (b) EL spectrum (green) and PL (black) spectrum of the OLET. Inset shows optical microscope image of the device. The EL emission (shown by arrow) is close to the edge of the Ca electrode. [*Source*: Namdas *et al.* [49]. Adapted with permission of AIP Publishing LLC.]

By implementing this measurement protocol, Namdas *et al.* reported brightness values exceeding 2500 cd/m^2 for an OLET where the active region is a bilayer film comprising a hole-transporting polymer, poly(2,5-bis)3-tetradecylthiophen-2-yl(thieno) [3,2-*b*] thiophene, and a polyphenylenevinylene-derivative light-emitting polymer, Super Yellow (Figure 6.6a). The optoelectronic saturation transfer characteristic shows that the field-effect charge transport is dominated by holes, given that the implemented charge-transport polymer does not present n-type transport, and Super Yellow is devoted to light generation. Thus, charge recombination (and consequent electroluminescence) takes place in the proximity of the drain electrode (Figure 6.6b) where the maximum of the brightness is achieved. As we have discussed in Chapter 4, when working in these bias conditions, the light formation process can be considered diode-like. The emission stripe width used for calculating the brightness value is 4.0 ± 0.5 μm.

It is straightforward that ambipolar conduction is crucial in light-emitting transistors for maximizing exciton recombination through electron–hole balance as well as for adjusting the position of the recombination region in the channel by tuning the gate voltage (thus, avoiding the exciton–electrode quenching). The consequent decrease in the exciton quenching will lead to improved brightness from the device.

In order to optimize the optoelectronic performances of bilayer-based polymeric OLET, a third top layer was inserted in order to enhance the electron injection.

Seo *et al.* [51] proposed to use a thin layer of a conjugated polyelectrolyte beneath the metal electrodes in the top contact/bottom gate bilayer OLET, whose active region is comprised by a hole-transport semiconductor and a light formation layer. In this way, the solution-processed conjugated polyelectrolyte layer most likely introduces ordered dipoles at the metal/organic semiconductor interface that facilitates electron injection. Indeed, this approach allowed the possibility of fabricating multicolor OLET by using different emissive layers but did not improve the achievable brightness values with respect the bilayer polymeric OLET. The reported brightness values are 137 cd/m^2 for blue, 647 cd/m^2 for green, and 112 cd/m^2 for red emission (Figure 6.7).

Until now, the multilayer approach showed the highest potential when implementing sublimated small molecules in the device active region [3], given the possibility of fine-tuning the buried interfaces. The achievement of very high EQE paves the way for obtaining intense brightness by the improvement of the charge-carrier mobility coupled to the implementation of triplet emitters. Indeed, the separation of light emission from the charge-carrier accumulation zones (thus, the separation of regions of large carrier densities from regions of large exciton densities) is also beneficial to intense light source realization.

Moreover, each layer is devoted to a single functionality within the device and can be optimized by controlling the growth of the different organic/organic, organic/metal, and organic/dielectric interfaces. Indeed, it is a very challenging task to synthesize a single material capable of providing both high current density and intense electroluminescence in single-layer field-effect devices, given that efficient charge transport and emission characteristics are typically self-excluding in well-ordered molecular organic systems.

Thus, other strategies can be implemented for improving brightness in OLETs such as engineering the overall device architecture for optimizing light-extraction and outcoupling as it is usually implemented in OLEDs.

Indeed, various approaches have been implemented to improve the external coupling of light in OLEDs by means of various internal and external device modification techniques such as substrate modification methods, use of scattering media, microlens arrays, microcavity, photonic crystals, nanocavities, nanoparticles, nanowires, nanostructures, and surface plasmon enhancing structures.

The specific features of the OLET architecture may guarantee unprecedented light extraction efficiencies by using ultrathin layers of ITO or Au as gate electrodes deposited on transparent substrates or by optimizing the thickness of the layers in the active regions. Moreover, it has to be noticed that in truly ambipolar OLET, light can be directly outcoupled from the emission zone to air without passing through any layer interface.

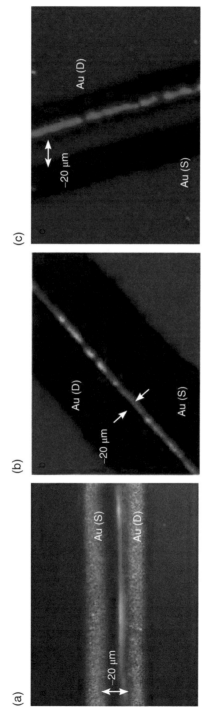

FIGURE 6.7 Photographs of operating (a) orange, (b) yellow-green, and (c) blue OLETs with electron-injecting conjugated polyelectrolyte layer. [*Source: Seo et al.* [51]. Adapted with permission from Wiley.]

The introduction of planar microcavities in OLEDs was demonstrated to modify the spontaneous emission by several studies [52–54]. The principal effect of the microcavity on the spontaneous emission is to redistribute the photon density of states such that only certain wavelengths, which correspond to allowed cavity modes, are emitted in a given direction. The spontaneous emission intensity is enhanced in the direction normal to the cavity axis relative to a noncavity device.

This approach can also be implemented in OLETs for improving the device brightness by inserting the multilayer cavity structure in the gate dielectric, which becomes in this way optically active [55]. Typically, the high-refractive-index and light-absorbing materials, such as doped Si and ITO, which are used to fabricate the gate electrode, introduce waveguiding effect and loss channels for the generated electroluminescence. Therefore, the interface between the dielectric and the active layer can be optically engineered by introducing a gate dielectric capable of reflecting light, which would be otherwise lost, to increase the device light output. Moreover, in addition to the photonic functionality, the optically active dielectric has to fulfill all the standard requirements for the gate dielectric of an OLET. Particularly, it needs to have the following: (i) low gate leakage current; (ii) the ability to sustain high voltage without dielectric breakdown; (iii) low density of electrons traps; (iv) high, or at least moderate, capacitance in order to have reasonable operating voltage; and (v) low surface roughness.

Namdas *et al.* [55] realized a gate dielectric with optical properties for enhancing the brightness of OLETs. The OLET emission wavelength is about four times the optical thickness of the multilayer dielectric. Thus, given the constructive interference of the reflections, the stack works as a high-quality reflector that considerably enhances the brightness of the emission from the device (Figure 6.8).

The optoelectronic gate dielectric acts as a standard gate dielectric that in addition redirects trapped emitted light into the forward direction, resulting in a substantial enhancement of the brightness of the OLET device to an extremely high value of $4500\,cd/m^2$.

As we have stated earlier, achieving balanced and intense electron and hole transport and hence efficient electron–hole recombination is particularly challenging. Unfortunately, ambipolar charge-transporting materials with sufficiently high mobility and PL emission efficiency are limited in number. Thus, a possible strategy to bypass this limitation is to engineer the device architecture in order to actively and independently control the injection, transport, and recombination of holes and electrons in the transistor channel.

As we anticipated in Chapter 3, Section 3.2, a split-gate OLET was realized to overcome the material constraint and simultaneously improve the

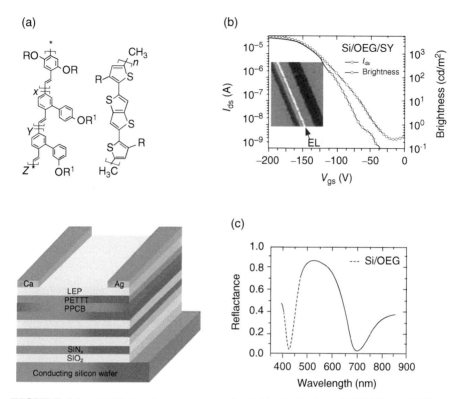

FIGURE 6.8 (a) Molecular structures of soluble derivative of PPV, Super-Yellow (SY) (chemical structure on the left) and poly(2,5-bis(3-tetradecylthiophen-2-yl) thieno[3,2-b]thiophene), (PBTTT) (chemical structure on the right), and lower panel is the schematic device architecture of OLET. The gate dielectric is formed on a substrate consisting of three pairs of alternating high- and low-refractive-index SiO_2/ SiN_x layers of quarter-wave thickness, which had been deposited on a heavily doped n-type silicon wafer. (b) Electrical transfer (square) and optical transfer (circle) characteristics of the multilayer-gate OLET. The gate voltage is scanned from 0 to -200 V while keeping the source–drain voltage at a fixed value of -200 V. The inset shows EL light emission photographed through a microscope. (c) Reflectance spectra of optoelectronic gate on silicon substrate. [*Source*: Namdas *et al.* [51]. Adapted with permission from Wiley.]

brightness and EQE [56]. A split-gate OLET has two independent gate electrodes that allow the device be operated as a unipolar transistor, a bipolar transistor, or a diode, in different I–V quadrants [57]. The control of charge injection by the two independent gate electrodes allows to enhance the transport of specific carrier species or switch off a specific current component. It is demonstrated that a maximum brightness of 609 cd/m^2 at 1.27% EQE can be achieved in a F8BT split-gate OLET, in which the recombination zone is confined at the center of the transistor channel.

For the sake of completeness, we mention that by optimizing the semiconductor layers thickness and collecting light through the gateless transparent substrate in OLETs based on the same active materials, calculated values of up to 8000 cd/m² are obtained [22]. Interestingly, this value is comparable to the best values achieved in fully optimized F8BT OLEDs [58].

The luminance values of OLETs are typically reduced by one order of magnitude when the entire device area, including the nonemitting parts of the channel and electrodes, is used for the calculation. The ratio between the lit and dark areas can be increased by engineering short channel length and small electrode pads. Because the brightness depends both on drain current and emission efficiency, the maximum brightness value is achieved at one edge of the ambipolar regime, where the recombination zone begins to detach from the electrode edge, and the EQE is about to reach the maximum in the ambipolar regime. At the point of lowest current in the middle of the ambipolar regime, the luminance drops to about half of the maximum value. Nonetheless, the OLET can work in a simultaneously very efficient and bright emission regime if one slightly moves in either direction out of this current minimum.

6.3   LIGHT OUTCOUPLING AND EMISSION DIRECTIONALITY

As we have seen in Section 6.1, the EQE of OLETs can be described by the general formula

$$\eta_{ext} = \eta_{int}\sigma_{out} \tag{6.3}$$

in addition to η_{int}, a parameter related to the fundamental optoelectronic properties of the active materials, one of the most important factors, which heavily affects both the brightness and the EQE, is the light outcoupling efficiency σ_{out}. σ_{out} determines the fraction of the photons generated within the device structure that can exit the device to reach the external open space.

The effective radiative quantum efficiency of the active layer in an OLET is derived from the intrinsic radiative quantum efficiency of the emitting material, which is reliably obtained if the emitting species is surrounded by an unbounded medium in the limit of very low excitation densities. Indeed, Φ_{PL} is determined experimentally by measuring a thin film of the emitter processed according to the characteristics of the active layer in the device, for example, a single-material layer or a host–guest dispersion.

By definition, Φ_{PL} is given by

$$\Phi_{PL} = \frac{\Gamma_R}{\Gamma_R + \Gamma_{NR}} \tag{6.4}$$

where Γ_R is the radiative decay rate of the excited state and Γ_{NR} is the sum of all competing nonradiative decay rates.

It is clear that every nonradiative contribution to the exciton decay reduces the radiative quantum efficiency. In addition, in organic optoelectronic devices such as OLEDs and multistack OLETs, the presence of stratified media with different refractive indices as well as the proximity to metallic electrodes may lead to a modification of the radiative decay rate of an exciton via the Purcell effect: $\Gamma_R \rightarrow \Gamma_R^* = F \cdot \Gamma_R$ (with F being the Purcell factor) [59–61].

By contrast, the nonradiative decay rates, for example, the dissipation of excitation energy into heat, are not affected by the cavity environment [61]. Thus, the effective radiative quantum efficiency can be defined as:

$$\Phi_{PL,eff} = \frac{F \cdot \Gamma_R}{F \cdot \Gamma_R + \Gamma_{NR}} \tag{6.5}$$

Depending on the details of the layer stack, the cavity effect can either increase or reduce $\Phi_{PL, eff}$ with respect to the intrinsic value Φ_{PL}. Consequently, in OLED engineering, the optimization of the cavity composed by the vertical stack structure is almost mandatory, not only in terms of the light outcoupling efficiency but also to enhance the radiative decay processes when Φ_{PL} is less than 1. In the case of OLETs, the situation is simplified by the inherent planar device geometry, in which the electrodes are microns away from the recombination zone, allowing light emission to be intrinsically cavity-free.

With reference to the previous discussion, we recall that even if $\Phi_{PL} = 1$, one has to be aware that in an operating device, the radiative quantum efficiency can be significantly reduced by exciton quenching mechanisms such as bimolecular exciton–exciton interaction or polaron–exciton quenching at high current densities [28] (see Chapter 5).

The internal quantum efficiency of organic optoelectronic devices can target the theoretical limit of 100%, if charge carrier injection and recombination are well balanced, if phosphorescent emitters are used, and if nonradiative exciton quenching processes are suppressed [62]. Nevertheless, even in the best ideal situation, only a fraction of the light generated in the device active region may be extracted to the outside space as a consequence of the refractive index mismatch among the different functional layers composing the device, such as the organic active layer/stack, the gate dielectric, the (transparent) substrate, the electrode(s), the encapsulating glass, and, finally, the ambient air.

In the case of the OLED platform, by assuming isotropic emission through a transparent anode and a perfectly reflecting cathode, the fraction of generated light escaping from the substrate can be evaluated according to a simple model based on ray optics by [63–66]:

$$\sigma_{out} = \frac{1}{\zeta n^2} \qquad (6.6)$$

where n denotes the (average) refractive index of the organic layer stack and ζ is a constant that depends on the dipole alignment and the geometry of the OLED device. With typical values of $n = 1.6$–1.8, it is easily found that only a fraction of 15–20% of the optical power is actually extracted from an OLED. However, this should only be taken as a rough estimation: a more sophisticated analysis has to take into account the coupling of the excited molecules to the modes of the OLED cavity.

From a general point of view, it has to be underlined that the OLET device architecture presents relevant advantages with respect to OLEDs not only in terms of control of the luminescence quenching induced by polarons and by metal electrodes but also in terms of light outcoupling.

Indeed in OLEDs, as we have seen, 80% of the emitted light is typically trapped in waveguided modes in the substrate and in the films. One major cause of this is the high refractive index of the transparent electrode [67], which cannot be removed and makes the design of efficient light outcoupling structures complex.

In an OLET, the generated light does not have to necessarily pass across a transparent conductor, which makes light outcoupling inherently more efficient. Simple optical structures, for instance, gratings, can be integrated into the OLET channel underneath the active layer without perturbing the lateral charge transport [68]. This might be particularly important for highly integrated devices where there is little space available for integrating light outcoupling structures.

Despite the importance of this issue, the optimization of the outcoupling efficiency of OLETs has not yet received adequate attention in the scientific community. For F8BT-based transistor in bottom contact and top gate configuration, an outcoupling efficiency through the gold gate electrode was derived to be as high as 16% by using a first-order approach based on a generic model for light extraction in OLEDs [6]. A comparative study for the trilayer heterostructure OLET resulted in a much higher outcoupling efficiency of around 27% when the gate electrode is made of ITO [3].

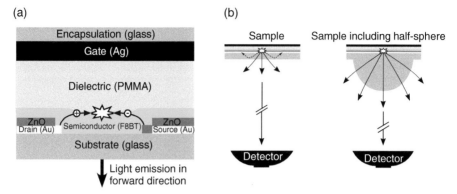

FIGURE 6.9 (a) Schematic of an encapsulated F8BT OLET emitting light in forward direction through the glass substrate. (b) Illustration of the standard measurement of the emitted light from the OLET sample using a photodetector positioned in the back of the sample in forward direction and a measurement configuration with an additional half-sphere mounted onto the substrate. [*Source*: Gwinner *et al.* [22]. Adapted with permission from Wiley.]

In order to increase the overall EQE in single-layer/top gate/bottom contact OLET, a study was performed in which the thickness of the polymeric active layer was kept constant (60 nm), while the thickness of the poly(methyl methacrylate) (PMMA) gate dielectric was varied between 460 and 620 nm [22]. Moreover, differently from a previously reported device structure, a 100 nm thick reflecting silver layer was used as gate electrode and light emission was measured from the back of the glass substrate (Figure 6.9)

Because of the reflecting silver gate, the measured electroluminescence emission of the OLETs collected through the glass substrate shows significant modifications induced by the different PMMA layer thickness (Table 6.1).

Simulated OLET spectra obtained from optical modeling of light emission and outcoupling (Figure 6.10b) are in agreement with the measurements and allow to indirectly determine the precise PMMA thicknesses. The angular emission pattern of the different OLETs deviates significantly from a standard Lambertian emission profile (Figure 6.10c). Again, the simulated results (Figure 6.10d) agree well with the experimental data, demonstrating the validity of the optical simulations.

Similar effects are usually observed in OLEDs, which can indeed be regarded as a one-dimensional microcavity, because the total thickness of the organic films in the device is typically of the same order of magnitude of the wavelength of the emitted light. There are two types of microcavities in OLED, that is, weak and strong microcavities [69]. Weak microcavity is formed with the conventional OLED structure due to the metal cathode and

TABLE 6.1   Summary of the Emission Characteristics and Performance Figures of the Ambipolar F8BT OLETs with Different PMMA Thicknesses.

PMMA Thickness (nm) (Bias V_4 (V))	Peak EQE (%)	Peak Luminance Efficiency (cd/A)	Peak Power Efficiency (lm/W)
460 (100)	8.16	28.20	0.89
530 (120)	7.62	23.63	0.62
560 (120)	6.00	28.61	0.49
620 (120)	6.38	21.82	0.57
460 (80)	7.69	26.57	1.04
460 (100) including half-sphere	12.46	42.87	1.37

Source: Gwinner *et al.* [22]. Adapted with permission from Wiley.

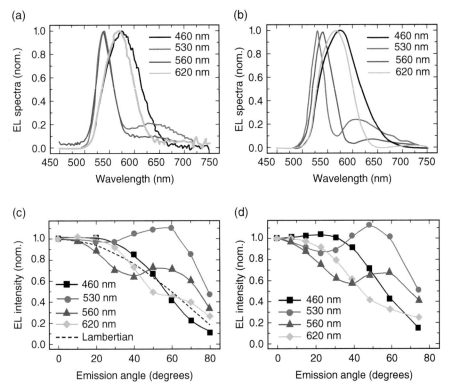

FIGURE 6.10   (a) Measured and (b) simulated EL spectra in forward direction for OLETs with different PMMA thicknesses. (c) Measured and (d) simulated angular emission pattern for OLETs with different PMMA thicknesses. [*Source*: Gwinner *et al.* [22]. Adapted with permission from Wiley.]

high-refractive-index anode (ITO) [70], while a strong microcavity structure usually is built by embedding a metal mirror on one side and a highly reflective dielectric multilayer structure on the other side of the organic stack [71]. According to this view, the conventional OLED structure acts as microcavity and weak reflections take place from the metal cathode and other reflective surfaces.

Clearly, weak microcavity effects are present when using reflecting silver gate in bottom contact/top gate single-layer OLET, even though they are less pronounced than in OLEDs given the absence of the high-refractive-index ITO.

The radiative lifetime of the molecular excited states can be strongly influenced by such microcavity. Lee *et al.* [70] studied the effect of microcavity on the light extraction efficiency of OLEDs. They have demonstrated that the substrate waveguided modes of an OLED structure strongly depend on the optical cavity length of the device due to weak cavity effect. It was found that the ratio of the extracted mode to the substrate-guided modes varies from 22% to 55%, depending on the location of recombination zone. They predicted that more light is trapped between the substrate and organic/ITO layers if the distance between the recombination zone and the cathode exceeds the quarter wavelength value.

Indeed, by modulating the thickness of the dielectric layer in the top gate/ bottom contact OLET device configuration, it is possible to maximize the flux of photons collected in the face-to-face detection scheme and, thus, the overall brightness and EQE.

A further enhancement in light outcoupling can be obtained by coupling a half-sphere with proper index-matching oil to the glass substrate of the device with the optimized dielectric layer thickness. Indeed, a spherically shaped substrate acts as a refractive-index-matching material as well as a lens so that the substrate waveguided modes are coupled out. Clearly, also the angular emission pattern of the optoelectronic devices is modulated by the presence of the half-sphere.

In bottom contact/top gate single-layer OLET, it was observed that the angle-integrated emission pattern is different from the Lambertian one and the correction factor is about 0.8. Moreover, it is claimed that the strongest feature introduced by the half-sphere is the suppression of the emission intensities for angles higher than 60°. The imperfect index matching and the interfacial unevenness at the substrate/oil/half-sphere interfaces manifest particularly at larger angles, leading to light losses within the plane reducing the outcoupling improvement. Despite this, the introduction of the half-sphere allowed to obtain a significant increase in the performance figures of the ambipolar F8BT OLETs, that is, 34% increase in EQE and peak luminance efficiency [22].

Differently from OLED technology, well-assessed methodology for measuring radiometric and photometric quantities is not common practice among researchers when working with OLETs. The protocol implemented by many authors for light collection consists in positioning a planar photodetector (10 mm diameter) in the forward direction in the proximity of the emitting source, and then correcting the collected electroluminescence signal for the angular pattern of the light dispersion with respect to the normal to the substrate in order to obtain the total emission intensity and EQE values. The correction factor is defined as the ratio of the angle integral of the actual emission pattern and the one obtained by assuming a Lambertian pattern. This method is used to mimic radiant intensity measurements without using an integrating sphere: this is a risky approach with multiple assumptions and should only be attempted when it is impossible to directly measure the relevant radiometric parameters.

At the early stage of the OLET development, direct photometric measurements were difficult due to the low-intensity emission and limited lifetime reliability of the fabricated devices. Numerical optical simulations of the emission outcoupling efficiency were reported in order to circumvent these experimental issues. Nonetheless, fundamental information about intrinsic photonic features of OLETs, together with useful hints for the engineering of the device structure for efficient light extraction, is provided by these pioneering works.

Gehlhaar *et al.* [72] used a numerical method, that is, finite-difference time-domain (FDTD) simulations, to calculate the interaction of electromagnetic waves with matter to perform an exhaustive theoretical study on the light extraction efficiency in the unipolar and ambipolar driving regime in OLETs. The considered device consisted in a stack of Si/dielectric/electrode/organic/electrode in bottom gate/bottom contact geometry. Particularly, they considered refractive index values of 1.45 and 1.7 for the dielectric (SiO_2) and the organic layers, respectively. Their simulations show that a maximum outcoupling efficiency of 30% is achievable in the case of a nonoptimized ambipolar OLET, regardless of the specific set of materials used in the device structure. In contrast, for a unipolar device, this model predicts a maximum outcoupling efficiency of 10% only and, therefore, affirms the advantage of the ambipolar design, also from a purely optical perspective.

Moreover, the use of highly reflective substrates and further reduction of the electrode absorption are shown to increase the device efficiency. All tested geometries show, however, a strong emission into guided modes, a loss channel that could only be tackled by surface modifications.

Particularly, by using gold electrode, the intensity of the guided modes is evidently damped, especially in the electrode direction. The use of an ITO layer as electrode and of an aluminum layer as reflector between the doped Si gate and the dielectric prompts the forward scattering of most of the emission and decreases the losses at the substrate. Therefore, the intensity of the extracted light is enhanced in comparison to the sample with metal electrodes (Figure 6.11).

Apart from studying the dependence of the extraction efficiency from the electrode layer nature and thickness, the authors reported plots of the modulation of the emission efficiency in different electrical driving regimes (unipolar and ambipolar). Particularly, they mimic the light emission processes in the different bias regimes by locating the emitters at different distances from the electrode edges. Indeed, it was demonstrated that increasing the distance from the electrode, the absorption of the emitted light by the metal decreases and the angle of the unaffected top emission increases. Moreover, an interference effect caused by reflections of light from the metal surface influences the spontaneous emission by changing the local electrical field.

All the calculated results are inherently correlated to single-layer devices with specific bottom gate and bottom contact device configuration and cannot be easily extended to other device architectures. Indeed, it was recently reported that in trilayer OLET in bottom gate and top contact configuration with ITO gate and polymeric PMMA dielectric, the angular emission profiles are almost Lambertian and invariant for unipolar and ambipolar bias regimes (Figure 6.12) [73].

In this specific case, the measurements of angular electroluminescence profiles were performed in order to experimentally investigate the vertical bulk distribution of excitons in the multilayer stack of the active region. It is well known that the spatial distribution and dipole orientation of the emitting excitons across the active layer(s) in organic optoelectronic devices can be inferred from electroluminescence emission profiles as it is typically done in OLEDs [74]. Surprisingly, the emission profiles obtained in the different bias regimes almost coincide despite the fact that, under unipolar conditions, a portion of EL is clearly emitted underneath the electrode.

In order to assess whether the lateral and vertical localization of the recombination zone along the multistack active region is expected to modify the angular light distribution, the measured OLET emission angular profiles were compared with the ones obtained by simulating light extraction from reference OLED-like multilayer structures. Indeed, light outcoupling in the ambipolar regime was simulated by a cathodeless structure formed by a glass/gate electrode/dielectric substrate and the same organic stack implemented in the trilayer OLET, that is, p-transport layer/recombination layer/n-transport layer. A top gold electrode is inserted in the simulated structure to reproduce the vertical architecture of the OLET in unipolar bias conditions. The radiative recombination of excitons on the emitting molecular sites in the emission layer is modeled by distributing isotropically oscillating dipoles at the interfaces

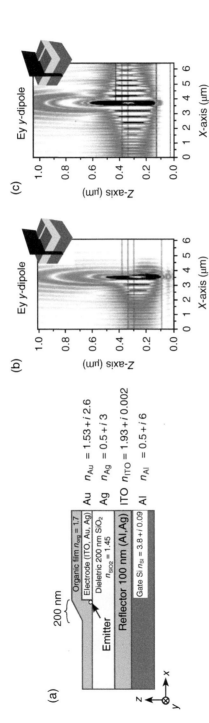

FIGURE 6.11 (a) Transistor structure as used in the FDTD simulation. (b, c) Comparison of electric field distributions of the light emission by a point source in OFETs with 40 nm thick electrodes of gold (b) and ITO (c). The vertically oriented electric field components of the corresponding plot plane are presented. For every projection, the plane of plot (black) is shown in an inset. [*Source:* Gehlhaar *et al.* [72]. Adapted with permission of AIP Publishing LLC.]

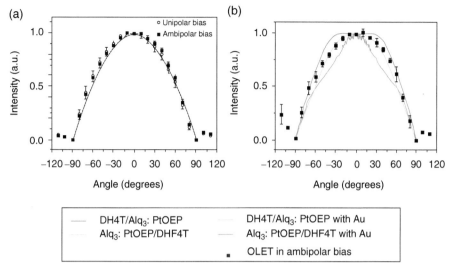

FIGURE 6.12 (a) EL emission angular profiles for a trilayer OLET with 1 mm width and 70 μm length, biased at $V_{ds} = V_{gs} = -100$ V (unipolar condition) and at $V_{ds} = -100$ V and $V_{gs} = -60$ V (ambipolar condition). (b) EL angular emission profile collected from the OLET in ambipolar condition (square symbols) compared with SETFOS-simulated profiles (solid lines) for device structures such as glass/ITO/PMMA/organic multilayer stack. The oscillating dipoles are located at the interface between the recombination and electron-transport layers (green and magenta lines) and between the hole-transport and the recombination layers (red and cyano lines). Finally, a top gold electrode is introduced on the organic multilayer stack (cyano and magenta lines). [*Source*: Toffanin *et al.* [73]. Adapted with permission from Wiley.]

between the recombination layer and the n-transport layer or between the recombination layer and the *p* transport layer.

As it can be observed in Figure 6.12b, the simulations of the vertical multilayer stack with the top gold electrode do not fit the experimental data for almost the entire angular range, regardless of the interface at which the recombination process is placed. Moreover, even though the simulations mimic the emitting zone of the OLET as a vertical stack without considering lateral structures, they suggest that the optical outcoupling in the devices is almost insensitive to the presence of the gold drain and source electrodes in all bias conditions, even when the emitting area extends for tens of microns under the electrode. Finally, since the experimental angular emission profiles lie in-between the simulated ones, which are calculated imposing exciton recombination at either the p-transport/recombination or the recombination/n-transport interface, it is suggested that the emission zone extends in the bulk of the recombination layer.

Considering the optical features of the OLETs reported in the previous examples, it is evident that light outcoupling optimization in OLETs is strictly correlated to the specific device architecture. Indeed, the implementation of either a single-layer or a multilayer as active region in the device is a distinguishing feature for controlling the efficiency of the devices, regardless of the applied bias. In the trilayer OLET device, the extracted-light spatial distribution is not affected by cavity effects and by the photon quenching process due to interaction with the source/drain electrode, even in unipolar bias conditions.

Moreover, the versatility of the OLET device architecture (such as bottom contact/top gate and bottom gate/top contact) allows for enhancing and easily optimizing the overall device outcoupling efficiency. Indeed, with respect to OLEDs, the outcoupling is more favorable in OLETs where (i) either no (transparent) electrode with high refractive index (such as ITO) causing severe in-plane waveguiding has to be passed by the emitted light or (ii) light can be extracted from the top side where no gate electrode and substrate are present. Indeed, the use of metal films as reflecting layer to be inserted between the dielectric and the gate substrate or even the use of a reflecting gate electrode is a simple method to recover the photon losses due to guided modes and absorption at the gate electrode.

Furthermore, the introduction of alternating materials forming the dielectric and a highly efficient distributed Bragg reflector is another option for eliminating loss channels in OLET devices. The reduction in electrode absorption by a replacement of metal films by conductive transparent layers is a further step toward efficient OLET light sources. However, conductive transparent films are not a simple alternative since absorption and conductivity are counter-correlated, and a low optical absorption leads to lower conductivity, a drawback for efficient electric microdevices.

The other significant source of photon losses is represented by the total internal reflection at the substrate/air interface [41]. As we have seen, there exist well-assessed methods to eliminate this photon-loss source, for example, by mounting an index-matched glass half-sphere with an index-matching oil onto the glass substrate.

Although it is not easy to implement all these requirements in real devices, it is evident that by careful engineering of the substrate as well as the device geometry, it is possible to achieve outstanding improvements in the optical performances of OLETs.

6.4   A POSSIBLE ROUTE FOR ORGANIC INJECTION LASING

OLET structural properties such as the in-plane geometry and the lateral charge injection are considered by many researchers as the key elements of a new strategy for the realization of the long-searched-for electrically pumped organic laser [5].

Optically pumped lasing over the entire range of the visible spectrum has been demonstrated using a number of polymeric and small-molecule organic semiconductors with different molecular structures. Electrically pumped lasers based on organic semiconductors possess numerous advantages over III–V semiconductor lasers, the most important being the higher integration potential on arbitrary substrates (including glass, polymers, and the backend of silicon complementary metal-oxide semiconductor (CMOS)), the lower cost, and the wider spectral range of possible lasing emission.

Despite success in optically pumped lasing, the electrically pumped organic laser remains a significant challenge [75]. Three major issues related to the organic device architecture must be settled in order to demonstrate an electrically pumped laser [76–78]: (i) the lasing threshold in devices containing injecting electrodes must be decreased; (ii) high carrier injection and high current densities must be demonstrated in the same device; (iii) optical losses from charge-induced absorption must be eliminated or strongly reduced.

Moreover, a significant breakthrough in materials design has to be carried out in order to noticeably reduce the annihilation of singlet and triplet excitons in the gain material, which is another major quenching mechanism preventing electrically pumped lasing [31].

Although the standard OLED configuration has been proposed, the losses in the two injecting electrodes overcome the potential gain in the organic semiconductor. In order to reduce the optical losses at the electrodes in the OLED structure, one strategy is to increase the thickness of the organic layers. However, because of the low carrier mobility in organic semiconductors, the thickness of the diode structure cannot be increased arbitrarily.

This makes it necessary to develop device architectures that are capable of reducing these negative effects to a minimum. The charge-transporting organic semiconductor should simultaneously exhibit high carrier mobility and high optical gain and needs to be incorporated into a low-loss resonator geometry.

As we have discussed in Chapter 5, the light-emitting transistor configuration allows moving the metal electrodes at a distance of micrometers from the region where exciton formation and light emission occur, thus preventing exciton quenching and photon losses due to interaction with the metal electrodes.

The high mobility of carriers moving in a field-effect configuration also allows minimization of the required charge-carrier densities in the device so as to limit exciton–charge quenching and photon absorption by polarons. Moreover, photonic structures for light guiding, confinement, and extraction can be easily incorporated in the OLET planar structure.

In ambipolar OLET devices, the only electrode that can introduce losses into the waveguide is the gate electrode that is separated from the gain region by the gate insulator. Implementing thin ITO film as gate electrode and low-refractive-index layers as gate insulator can assure efficient amplified spontaneous emission (ASE), which is the necessary condition for lasing action to take place [79].

As a general statement, lowering ASE threshold diminishes the current density required to achieve electrically pumped lasing emission and reduces the polaron-induced absorption in the "gain" medium [80]. Therefore, great efforts are devoted to synthesizing new materials and to engineering new device structures with lower ASE threshold and enhanced net gain coefficient. In this way, it would be possible to achieve lasing emission in real devices at an achievable current density even in the presence of residual exciton quenching and photon losses. Very low ASE threshold and high PL quantum efficiency values are demonstrated in different organic conjugated systems presenting both field-effect and electroluminescence properties (single-component small-molecule [81] and polymeric [82] thin film and dilute solution of dyes embedded in host matrices [83]).

Among the various photonic structures that can be utilized to provide resonant feedback in organic semiconductor lasers, distributed feedback (DFB) geometries are worthy of particular attention as they offer high reflectivities, long gain lengths, and high optical confinement of the oscillating mode, leading to very low operation thresholds [75].

In addition, such structures offer high output wavelength purity and stability, and their fabrication is compatible with in-plane OLET geometry. Furthermore, DFB structures can be fabricated by either standard lithographic techniques or by simpler soft lithography methods (such as micromolding and embossing) [84–86].

Namdas et al. [87] demonstrated that one-dimensional and two-dimensional DFB gratings can be directly incorporated into polymer-based OLET device structure by nanoimprint lithography. Optically pumped lasing was achieved from a DFB-nanoimprinted Super-Yellow polymer film, drop cast onto a substrate comprising a gate dielectric spin-coated onto the gate conducting (ITO) electrode, which mimics losses from the OLET architecture. The authors estimated that the threshold current density necessary for electrically pumped lasing is a factor 50 higher than the current density in their optimized polymer-based OLETs. Different approaches can be implemented to enhance the current density, such as reducing the channel length, increasing the gate capacitance, using polymers with higher PL quantum efficiency, and using pulsed operation to achieve higher voltages.

Gwinner et al. [88] moved further toward the complete integration of a photonic structure into a fully working OLET. They reported on a rib

(a)

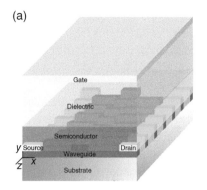

(b)

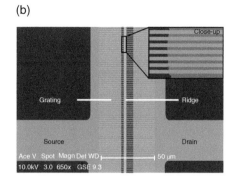

FIGURE 6.13 (a) Schematic illustration of an F8BT-based top gate/bottom contact OLET structure with integrated waveguide rib and DFB grating. (b) Environmental scanning electron microscope image of the fabricated Ta_2O_5 waveguide rib structure with additional DFB grating aligned to the T-shaped gold electrode pattern. The inset shows a close-up of the grating on top and next to the ridge. [*Source*: Gwinner *et al.* [88]. Adapted with permission from Wiley.]

waveguide DFB structure consisting of tantalum pentoxide (Ta_2O_5) integrated within the channel of a top gate/bottom contact F8BT-based OLET (Figure 6.13). The emitted light is coupled efficiently into the resonant mode of the DFB waveguide when the recombination zone of the OLET is placed directly above the waveguide rib. Indeed, moving the recombination zone within the channel by the applied voltages modifies the EL spectrum of the device. Through a combination of strong two-dimensional confinement in the rib waveguide structure, the use of a silver gate electrode, and an optimized gate dielectric thickness, optical losses at metal electrodes are eliminated.

Unfortunately, the achievable electrical pumping power calculated for the best-performing F8BT-based OLETs is four orders of magnitude below the singlet exciton density necessary to achieve electrically pumped lasing. Moreover, additional losses encountered in electrically pumped structures, such as charge-induced absorption and triplet-induced losses, are neglected in this calculation. Clearly, all these estimations on the threshold current density value are strictly dependent on the assumptions made, which concern the width and thickness of the charge-accumulation layer.

Indeed, it was demonstrated that the F8BT postannealing treatment that optimizes the ambipolar transistor current, and thus light emission intensities, is detrimental for waveguiding and lasing processes [89]. An efficient way of tuning ambipolar charge-carrier injection and improving the electrode/organic interface morphology is to modify the gold electrodes with different types of self-assembled monolayers. However, the effects of interface modifications are yet to be thoroughly analyzed.

All of this highlights how difficult and challenging it is to simultaneously optimize electronic, optical, and lasing properties in complex multifunctional optoelectronic organic devices.

Apart from its potential in the context of lasers, waveguide structures incorporated in OLET architecture are fundamental for optoelectronic integrated circuits, where a light signal generated by an OLET integrated with a transmitter circuit could be coupled efficiently into a low-loss waveguide toward a detection device coupled to a receiver circuit. Shigee and coworkers [90] have proposed a simpler, though less effective, photonic device structure in order to confine the scattered-out light from the line-shaped recombination zone in OLETs. In this study, a unipolar small-molecule semiconductor layer was deposited onto a channel-shaped gate insulator waveguide of SiO_2, and symmetric source/drain electrodes of gold were formed on the side walls of the waveguide by self-shadow-masking vapor deposition.

Under AC gate operation, light emission was achieved due to recombination of injected electrons with the accumulated holes at high-frequency gate voltage cycles. However, the observed light intensity was low due to less-effective electron injection at the high-work-function gold electrode.

As we have already discussed, in single-layer ambipolar OLETs [15,91], light emission is vertical to the substrate plane with the emission zone being a line parallel to the source and drain electrodes. By contrast, it has been demonstrated that in single-crystal OLETs, the EL emission is strongly self-waveguided at both edges of the crystal with respect to the almost absent surface emission [25,92]. Indeed, the light generated with small grazing angle is totally internally reflected at the two slab interfaces [25].

As we reported in Chapter 3, Bisri et al. [25] realized an OLET based on BP3T single crystal where the device active region acts as a waveguide structure. This device structure gives the possibility of collecting all of the generated light at very narrow single (or double) exits, which coincide with the crystal edges. Moreover, this device structure allows to achieve spectral narrowing of the main EL peak when high drain current is achieved in the ambipolar regime. Particularly, it was observed that devices with wider channel tended to yield a higher photon density from the edge (Figure 6.14a). Two main mechanisms seem reasonable to explain the spectral narrowing: narrowing via cut-off mode [93], in which a mode is selected as it propagates along the organic/substrate interface due to perfect reflection; narrowing via ASE due to the self-waveguiding of the emission and the gain properties of the organic layer. Neither of the two described mechanisms is conclusive for explaining the experimental results.

(a)

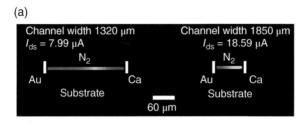

(b)

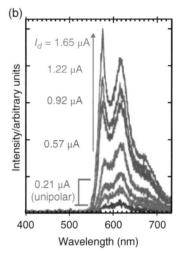

FIGURE 6.14 (a) Light-confinement effect on the edge-emission light density. Side views of the slab-stacked BP3T single-crystal LETs are shown. The device structure confines the light without any emission from the crystal surface and makes all of the produced light propagate only to the crystal edge. Wider channel gives denser edge emission due to light confinement, in addition to the slight operating-current increase caused by the short channel length. (b) Drain current-dependent spectral evolution, measured in real time during a drain voltage sweep with consequent spectral narrowing at a high current regime. Brighter emission is observed from the device that has a channel length of 400 μm. [*Source*: Bisri *et al*. [25]. Adapted with permission from Wiley.]

We have to mention that Yamao *et al*. [94] also achieved the spectrally narrowed emission using a crystal of 2,5-bis(4-biphenylyl)thiophene (BP1T). They observed a narrowed spectral line at 491.5 nm with FWHM ~1.1 nm. However, in contrast with the results of Bisri *et al*., Yamao *et al*. did not observe any spectral evolution, probably due to the high operation voltages of the device.

A novel OLET device has been developed [92] that includes a diffraction grating integrated in the SiO_2 gate insulator, but outside the device channel.

Once the device is operated by applying square-wave alternating gate voltages, EL is waveguided in the slab crystal and finally coupled into the grating. As a result, the spectral narrowing emission is obtained under relatively low voltages as a dominant line with 2.05 nm FWHM. Thus, the diffraction grating functions as a spectral filter, somehow resembling a Fabry–Pérot etalon [95].

The current-induced spectral narrowing found in these experiments makes ambipolar OLETs based on organic single crystals the best candidates to obtain electrically driven laser oscillations from organic materials, once high and well-balanced values of electron and hole mobility are achieved.

6.5 CONCLUSIONS

The photonic properties of OLETs are the essential features that might enable or hinder any practical applications of this new technology. The brightness, which is strictly entangled with the internal emission quantum efficiency and the outcoupling efficiency, has to be evaluated considering the specific geometrical characteristics of the device structure. Indeed, the device structure is the most distinguishing feature of OLETs, which offers an alternative platform with respect to OLEDs for achieving high efficient and bright electroluminescence from organic materials. The thorough understanding of the specificity of the OLET structure and properties is key for identifying the direction of any technological developments. Noteworthy, in OLETs, it is possible to achieve simultaneous high EQE and brightness, which enable a range of applications when coupled to high power efficiency, low driving voltage, and reliability. In the next chapter, we will focus our attention on the most promising applications of this new technological paradigm.

REFERENCES

[1] R. H. Friend, R. W. Gymer, A. B. Holmes, J. H. Burroughes, R. N. Marks, C. Taliani, D. D. C. Bradley, D. A. Dos Santos, J. L. BreÂ das, M. Logdlund, and W. R. Salaneck, *Nature* **1999**, *397*, 12.

[2] J. Cornil, J. L. Bredas, J. Zaumseil, and H. Sirringhus, *Adv. Mater.* **2007**, *19*, 1791.

[3] R. Capelli, S. Toffanin, G. Generali, H. Usta, A. Facchetti, and M. Muccini, *Nature Mater.* **2010**, *9*, 496.

[4] M. Pope, and C. E. Swenberg, *Electronic Processes in Organic Crystals and Polymers*, 1st edn. (Oxford University Press, Oxford, **1982**), 1328.

[5] M. A. Baldo, R. J. Holmes, and S. R. Forrest, *Phys. Rev. B* **2002**, *66*, 035321.

[6] J. Zaumseil, C. R. McNeil, M. Bird, D. L. Smith, P. P. Ruden, M. Roberts, M. J. McKiernan, R. H. Friend, and H. Sirringhaus, *J. Appl. Phys.* **2008**, *103*, 064517.

[7] H. Yamamoto, T. Oyamada, H. Sasabe, and C. Adachi, *Appl. Phys. Lett.* **2004**, *84*, 1401.

[8] S. Verlaak, D. Cheyns, M. Debucquoy, V. Arkhipov, and P. Heremans, *Appl. Phys. Lett.* **2004**, *85*, 2405.

[9] S. Haneder, E. Da Como, J. Fedmann, J. M. Lupton, C. Lennartz, P. Erk, E. Fuchs, O. Molt, I. Munster, and W. G. Schildknecht, *Adv. Mater.* **2008**, *20*, 3325.

[10] J. Kalinowski, H. Murata, L. C. Picciolo, and Z. H. Kafafi, *J. Phys. D* **2001**, *34*, 21.

[11] W. Stampor, J. Kalinowski, P. Di Marco, and V. Fattori *Appl. Phys. Lett.* **1997**, *70*, 1935.

[12] S. Inal, M. Schubert, A. Sellinger, and D. J. Neher, *J. Phys. Chem. Lett.* **2010**, *1*, 982.

[13] J. Kalinowski, J. Mezyk, F. Meinardi, R. Tubino, M. Cocchi, and D. Virgili, *J. Chem. Phys.* **2008**, *128*, 124712.

[14] J. Lloyd-Hughes, L. M. Herz, M. B. Johnston, T. Richards, and H. Sirringhaus, *Phys. Rev. B* **2008**, *77*, 125203.

[15] J. Zaumseil, C. L. Donley, J.-S. Kim, R. H. Friend, and H. Sirringhaus, *Adv. Mater.* **2006**, *18*, 2708.

[16] T. Takenobu, S. Birsi, T. Takahashi, M. Yahiro, C. Adachi, and Y. Iwasa, *Phys. Rev. Lett.* **2008**, *100*, 15.

[17] C. Tanase, E. J. Meijer, P. W. M. Blom, and D. M. de Leeuw, *Org. Electron.* **2003**, *4*, 33.

[18] G. Paasch, T. Lindner, C. Rost-Bietsch, S. Karg, W. Riess, and S. Scheinert, *J. Appl. Phys.* **2005**, *98*, 084505.

[19] D. L. Smith and P. P. Ruden, *J. Appl. Phys.* **2007**, *101*, 084503.

[20] C. T. Hsieh, D. S. Citrin, and P. P. Ruden, *Appl. Phys. Lett.* **2007**, *90*, 012118.

[21] M. H. Lu and J. C. Sturm, *J. Appl. Phys.* **2002**, *91*, 595.

[22] M. C. Gwinner, D. Kabra, M. Roberts, T. J. K. Brenner, B. H. Wallikewitz, C. R. McNeill, R. H. Friend, and H. Sirringhaus, *Adv. Mater.* **2012**, *24*, 2728.

[23] Z. Chen, M. J. Lee, R. S. Ashraf, Y. Gu, S. Albert-Seifried, M. M. Nielsen, B. Schroeder, T. D. Anthopoulos, M. Heeney, I. McCulloch, and H. Sirringhaus, *Adv. Mater.* **2012**, *24*, 647.

[24] M. Melucci, M. Zambianchi, L. Favaretto, M. Gazzano, A. Zanelli, M. Monari, R. Capelli, S. Troisi, S. Toffanin, and M. Muccini, *Chem. Commun.* **2011**, *47*, 11840.

[25] S. Z. Bisri, T. Takenobu, Y. Yomogida, H. Shimotani, T. Yamao, S. Hotta, and Y. Iwasa, *Adv. Funct. Mater.* **2009**, *19*, 1728.

[26] K. Sawabe, T. Takenobu, S. Z. Bisri, T. Yamao, S. Hotta, and Y. Iwasa, *Appl. Phys. Lett.* **2010**, *97*, 043307.

[27] K. Sawabe, M. Imakawa, M. Nakano, T. Yamao, S. Hotta, Y. Iwasa, and T. Takenobu, *Adv. Mater.* **2012**, *24*, 6141.

[28] N. C. Giebink and S. R. Forrest, *Phys. Rev. B* **2008**, *77*, 235215.

[29] F. So, J. Kido, and P. Burrow, *MRS Bull.* **2008**, *33*, 663.

[30] L. S. Hung and C. H. Chen, *Mater. Sci. Eng. R: Rep.* **2002**, *3*, 143

[31] N. C. Giebink and S. R. Forrest, *Phys. Rev. B* **2009**, *79*, 073302.

[32] M. Lehnhardt, T. Riedl, T. Weimann, and W. Kowalsky, *Phys. Rev. B* **2010**, *81*, 165206.

[33] H. Nakanotani, T. Oyamada, Y. Kawamura, H. Sasabe, and C. Adachi, *Jpn. J. Appl. Phys., Part 1*, **2005**, *44*, 3659.

[34] M. A. Baldo, D. F. O'Brien, Y. You, A. Shoustikov, S. Sibley, M. E. Thompson, and S. R. Forrest, *Nature*, **1998**, *395*, 151.

[35] K. Goushi, R. Kwong, J. J. Brown, H. Sasabe, and C. Adachi, *J. Appl. Phys.* **2004**, *95*, 7798.

[36] Y. Sun, N. C. Giebink, H. Kanno, B. Ma, M. Thompson, and S. Forrest, *Nature* **2006**, *440*, 908.

[37] G. Schwartz, K. Fehse, M. Pfeiffer, K. Walzer, and K. Leo, *Appl. Phys. Lett.* **2006**, *89*, 083509.

[38] D. Qin and Y. Tao, *Appl. Phys. Lett.* **2005**, *86*, 113507.

[39] G. Schwartz, M. Pfeiffer, S. Reineke, K. Walzer, and K. Leo, *Adv. Mater.* **2007**, *19*, 3672.

[40] N. C. Greenham, R. H. Friend, and D. D. C. Bradley, *Adv. Mater.* **1994**, *6*, 491.

[41] S. Reineke, F. Lindner, G. Schwartz, N. Seidler, K. Walzer, B. Lussem, and K. Leo, *Nature* **2009**, *459*, 234.

[42] J. S. Swensen, J. Yuen, D. Gargas, S. K. Buratto, and A. J. Heeger, *J. Appl. Phys.* **2007**, *102*, 013103.

[43] J. Zaumseil, R. H. Friend, and H. Sirringhaus, *Nat. Mater.* **2006**, *5*, 69.

[44] E. B. Namdas, P. Ledochowitsch, J. S. Swensen, J. D. Yuen, D. Moses, and A. J. Heeger, *Adv. Mater.* **2008**, *20*, 1321.

[45] A. Hepp, H. Heil, W. Weise, M. Ahles, R. Schmechel, and H. von Seggern, *Phys. Rev. Lett.* **2003**, *91*, 157406.

[46] T. Takahashi, T. Takenobu, J. Takeya, and Y. Iwasa, *Adv. Funct. Mater.* **2007**, *17*, 1623.

[47] K. Nakamura, M. Ichikawa, R. Fushiki, T. Kamikawa, M. Inoue, T. Koyama, and Y. Taniguchi, *Jpn. J. Appl. Phys., Part 1* **2005**, *44*, L1367.

[48] *Handbook of Applied Photometry* Ed. C. DeCusatis (Optical Society of America and Springer-Verlag New York, Inc., **1998**).

[49] E. B. Namdas, P. Ledochowitsch, J. D. Yuen, D. Moses, and A. J. Heeger, *Appl. Phys. Lett.* **2008**, *92*, 183304.

[50] Y. Shao, G. C. Bazan, and A. J. Heeger, *Adv. Mater.* **2007**, *19*, 365.

[51] J. H. Seo, E. B. Namdas, A. Gutacker, A. J. Heeger, and G. C. Bazan *Adv. Funct. Mater.* **2011**, *21*, 3667.

[52] A. Dodabalapur, L. J. Rothberg, and T. M. Miller, *Appl. Phys. Lett.* **1994**, 2308.

[53] U. Lemmer, R. Hennig, W. Guss, A. Ochse, J. Pommerehne, R. Sander, A. Greiner, R. F. Mahrt, H. Bassler, J. Feldmann, and E. O. Gobel, *Appl. Phys. Lett.* **1995**, *66*, 1301.

[54] H. F. Wittmann, J. Gruner, R. H. Friend, G. W. C. Spencer, S. Moratti, and A. B. Holmes, *Adv. Mater.* **1995**, *7*, 541.

[55] E. B. Namdas, B. B.Y. Hsu, J. D. Yuen, I. D. W. Samuel, and A. J. Heeger, *Adv. Mater.* **2011**, *23*, 2353.

[56] B. B.Y. Hsu, C. Duan, E. B. Namdas, A. Gutacker, J. D. Yuen, F. Huang, Y. Cao, G. C. Bazan, I. D. W. Samuel, and A. J. Heeger *Adv. Mater.* **2012**, *24*, 1171.

[57] B. B. Y. Hsu, E. B. Namdas, J. Yuen, I. D. W. Samuel, and A. J. Heeger, *Adv. Mater.* **2010**, *22*, 4649.

[58] D. Kabra, L. P. Lu, M. H. Song, H. J. Snaith, and R. H. Friend, *Adv. Mater.* **2010**, *22*, 3194.

[59] B. C. Krummacher, S. Nowy, J. Frischeisen, M. Klein, and W. Brutting, *Org. Electron.* **2009**, *10*(3), 478.

[60] S. Nowy, B. C. Krummacher, J. Frischeisen, N. A. Reinke, and W. Brutting, *J. Appl. Phys.* **2008**, *104*(12), 123109. *Phys. Status Solidi A 2013*, 210, 63.

[61] H. Becker, S. E. Burns, and R. H. Friend, *Phys. Rev. B* **1997**, *56*(4), 1893.

[62] J. S. Kim, P. K. H. Ho, N. C. Greenham, and R. H. Friend, *J. Appl. Phys.* **2000**, *88*, 1073.

[63] S.R. Forrest, D.D.C. Bradley, and M.E. Thomson, *Adv. Mater.* **2003**, *15*, 1043.

[64] N. K. Patel, S. Cinà, and J. H. Burroughes, *IEEE J. Select. Top. Quant. Electron.* **2002**, *8*, 346.

[65] D.S. Mehta and K. Saxena, *Proceedings of the Ninth Asian Symposium on Information Display (ASID-06)*, **2006**, 198.

[66] S. R. Forrest, *Org. Electron.* **2006**, *4*, 45.

[67] K. Saxena, V. K. Jain, D. S. Mehta, *Opt. Mater.* **2009**, *32*, 221.

[68] M. Muccini, W. Koopman, and S. Toffanin, *Laser Photonics Rev.* **2012**, *6*, 258.

[69] V. Bulovic, V. B. Khalfin, G. Gu, P. E. Burrows, D. Z. Garbuzov, and S. R. Forrest, *Phys. Rev. B* **1998**, *58*, 3730.

[70] J. Lee, N. Chopra, and F. So, *Appl. Phys. Lett.* **2008**, *92*, 033303.

[71] E. F. Schubert, N. E. J. Hunt, M. Micovic, R. J. Malik, D. L. Sivco, A. Y. Cho, and G. J. Zydzik, *Science* **1994**, *265*, 943.

[72] R. Gehlhaar, M. Yahiro, and C. Adachi, *J. Appl. Phys.* **2008**, *104*, 033116.

[73] S. Toffanin, R. Capelli, W. Koopman, G. Generali, S. Cavallini, A. Stefani, D. Saguatti, G. Ruani, and M. Muccini, *Laser Photonics Rev.* **2013**, *7*, 1011.

[74] S. L. M. van Mensfoort, M. Carvelli, M. Megens, D. Wehenkel, M. Bartyzel, H. Greiner, R. A. J. Janssen, and R. Coehoorn, *Nature Phot.* **2010**, *4*, 329.

[75] I. D. W. Samuel, E. B. Namdas, and G. A. Turnbull, *Nature Photon.* **2009**, *3*, 546.

[76] M. Reufer, S. Riechel, J. M. Lupton, J. Feldmann, U. Lemmer, D. Schneider, T. Benstem, T. Dobbertin, W. Kowalsky, A. Gombert, K. Foberich, V. Wittwer, and U. Scherf, *Appl. Phys. Lett.* **2004**, *84*, 3262.

[77] P. Gorm, T. Rabe, T. Riedl, and W. Kowalsky, *Appl. Phys. Lett.* **2007**, *91*, 041113.

[78] S. Lattante, F. Romano, A. P. Cariato, M. Martino, and M. Anni, *Appl. Phys. Lett.* **2006**, *89*, 031108.

[79] M. Pauchard, J. Swensen, D. Moses, A. J. Heeger, E. Perzon, and M. R. Andersson, *J. Appl. Phys.* **2003**, *94*, 3543.

[80] H. Yamamoto, H. Kasajima, H. Sasabe, and C. Adachi, *Appl. Phys. Lett.* **2005**, *86*, 83502.

[81] H. Nakanotani, S. Akiyama, D. Ohnishi, M. Moriwake, M. Yahiro, T. Yoshihara, S. Tobita, and C. Adachi, *Adv. Funct. Mater.* **2007**, *17*, 2328.

[82] R. Xia, G. Heliotis, P. N. Stavrinou, and D. D. C. Bradley, *Appl. Phys. Lett.* **2005**, *87*, 031104.

[83] S. Toffanin, R. Capelli, T. Y. Hwu, K. T. Wong, T. Plotzing, M. Forst, and M. Muccini, *J. Phys. Chem. B* **2010**, *114*, 120.

[84] D. Pisignano, M. Anni, G. Gigli, R. Cingolani, G. Barbarella, L. Favaretto, and G. Sotgiu, *Synth. Met.* **2003**, *137*, 1057.

[85] M. Breggren, A. Dodabalapur, R. E. Slusher, A. Timko, and O. Nalamasu, *Appl. Phys. Lett.* **1999**, *74*, 3257.

[86] J. A. Rogers, M. Meier, A. Dodabalapur, E. J. Laskowski, and M. A. Cappuzzo, *Appl. Phys. Lett.* **1999**, *74*, 3257.

[87] E. B. Namdas, M. Tong, P. Ledochowitsch, S. R. Mednick, J. D. Yuen, D. Moses, and A. J. Heeger, *Adv. Mater.* **2009**, *21*, 799.

[88] M. C. Gwinner, S. Khodabakhsh, M. H. Song, H. Schweizer, H. Giessen, and H. Sirringhaus, *Adv. Funct. Mater.* **2009**, *19*, 1.

[89] M. C. Gwinner, S. Khodabakhsh, H. Giessen, and H. Sirringhaus, *Chem. Mater.* **2009**, *21*, 4425.

[90] Y. Shigee, H. Yanagi, K. Terasaki, T. Yamao, and S. Hotta, *Jpn. J. Appl. Phys.* **2010**, *49*, 01AB09.

[91] J. S. Swensen, C. Soci, and A. J. Heeger, *Appl. Phys. Lett.* **2005**, *87*, 253511.

[92] T. Yamao, Y. Sakurai, K. Terasaki, Y. Shimizu, H. Jinnai, and S. Hotta, *Adv. Mater.* **2010**, *22*, 3708.

[93] D. Yokoyama, M. Moriwake, and C. Adachi, *J. Appl. Phys.* **2008**, *103*, 104.

[94] T. Yamao, K. Terasaki, Y. Shimizu, and S. Hotta, *J. Nanosci. Nanotechnol.* **2010**, *10*, 1017.

[95] F. G. Smith, T. A. King, and D. Wilkins, *Optics and Photonics: An Introduction*, 2nd edn. (Wiley, Chichester, UK, **2007**), 198.

7

APPLICATIONS OF ORGANIC LIGHT-EMITTING TRANSISTORS

Organic-based optoelectronic devices hold a great promise of key advantages over the existing technologies, including simplicity of construction and potential cost savings. Therefore, there is growing interest in the development and use of all-organic components and systems for the existing and radically new applications. Apart from the obvious applications of displays and general lighting, organic light sources are well suitable for optical sensing [1], on-chip spectroscopy [2], and data communications [3,4].

Organic light-emitting transistors (OLETs) show a versatile and attractive architecture for low-loss light-signal transmission in optoelectronic integrated circuits. To date, frequency-modulated gate voltages are mainly applied only to study the carrier-injection and -transport mechanisms in active organic semiconducting materials [5,6]. However, the multifunctionality intimately inherent within the OLET platform can lead to new applications in optoelectronic devices. Indeed, recently, a proof-of-principle device has been realized that combines the emission properties of an OLET with the electrical properties of an inverter and thereby functions as an electrically driven optoelectronic switch, in which high and low states can be read both optically and electrically [7].

Given the recent results in the fabrication of bright, efficient, and reliable devices, it is expected in the near future that the full compatibility of field-effect light-emitting devices with well-established electronic and photonic planar technologies will allow the development of viable technological

Organic Light-Emitting Transistors: Towards the Next Generation Display Technology, First Edition. Michele Muccini and Stefano Toffanin.
© 2016 John Wiley & Sons, Inc. Published 2016 by John Wiley & Sons, Inc.

solutions in various application fields, including display technology and sensing. Particularly, OLETs may constitute a key element for the development of next-generation organic active-matrix display technology. The increase in pixel brightness and lifetime due to an unparalleled control of charge injection and accumulation in the organic layer, the combination in a single device of the electrical-switching and light-emission functionalities, which reduces the number of switching thin-film transistors (TFTs) to be realized in the active-matrix driving circuit, and the gate voltage modulation of the light emission, which enables the use of lower quality TFT backplanes, make OLETs a potentially breakthrough technology for display applications.

In addition, the intrinsic electrical amplification of the field-effect device makes OLETs particularly suitable for highly efficient sensing. The sensing principle may be built on the current modulation, and consequent light emission modulation, induced by the interaction of analytes with the active interface of the device. Alternatively, the OLET can be used as a miniaturized low-power light-emitting source integrated into a sensing system with other photonic and electronic components.

In this chapter, we discuss the distinguishing features of OLET display and sensing applications and highlight the most relevant open issues and next development targets.

7.1 OLET DISPLAY TECHNOLOGY

Today's flat-panel display (FPD) industry is dominated by liquid-crystal displays (LCDs), particularly in its active-matrix liquid-crystal displays (AM-LCDs) form. Several other FPD technologies (e.g., field-emission or electroluminescent (EL) displays), which trace their origins at least as far back as LCDs, have either fallen by the wayside or, at best, penetrated for a limited period of time only limited display market segments. Concerning organic light-emitting diodes, a big question arose and persists. Will active-matrix organic light-emitting diodes displays (AM-OLEDs) fare any better against AM-LCDs, which today have become an entrenched FPD technology with a large installed manufacturing base and well-developed infrastructure for parts supply and integration? It is a generally accepted statement that the device geometry of organic light-emitting diodes (OLEDs) is virtually ideal for FPD applications: a layer of luminescent material is placed between two electrodes in a sandwich geometry; light is emitted on the passage of electrical current supplied by the active-matrix pixel electrode circuits, with a color that, in general, depends on the choice of organic materials. It has been considered virtually impossible to imagine anything simpler, flatter, or lighter.

More specifically, the following OLED characteristics are considered winning features for FPD applications: Lambertian self-emission [8], which produces a wide viewing angle; fast response time (below microseconds), which is suitable for video rate; high luminous efficiency and low operation voltage, which guarantee low power consumption; lightweight, very thin structure, and robustness to the external impact as well as good daylight visibility through high brightness and contrast, which are desirable for portable displays; broad color gamut, which is suitable for full-color displays; simple, low-temperature fabrication processes and low-cost organic materials, which make manufacturing cost-effective; and thin-film conformability on plastic substrates. All of these properties render in principle OLEDs a promising candidate for future flexible display applications [9,10]. However, when turning to practical implementation, many big issues are encountered, which represent significant obstacles to be overcome before the potential of this new technology can be realized. These include integration with the drive electronics, fine patterning of the multicolor frontplane, efficient light output extraction; power efficiency and consumption, full cost analysis of manufacturing, and when looking at nonrigid applications, flexible barriers to guarantee the needed operating and storage lifetime.

To solve some of these important issues, tremendous research efforts from academia and industry have been underway since the first appearance in 1997 of the OLED-based monochromatic car stereo displays in the market [11]. AM-OLED displays based on TFT backplane technology, in particular, have attracted considerable attention for high-resolution, medium- to large-size FPD applications, such as portable devices, laptops, and TV screens. Indeed, AM-OLED displays have lots of advantages for FPD, such as fast response time, high contrast ratio, wide viewing angle, and thinness [12]. Concerning the backplane, low-temperature polysilicon (LTPS) TFTs using excimer laser annealing (ELA), solid-phase crystallization (SPC), and sequential lateral solidification (SLS) have been widely used as the pixel element in AM-OLED because of their excellent current driving ability and electrical reliability [13]. Polycrystalline silicon (poly-Si) TFT technology [14] has better electrical performance and operational stability, at the expenses of a much higher cost, in comparison to the much cheaper hydrogenated amorphous silicon (a-Si:H) TFTs. Since a much matured a-Si:H TFT technology is already available at low cost from the AM-LCD industry, there have been a long-time hope that the a-Si:H TFT technology might challenge the poly-Si TFT technology for use in AM-OLEDs. However, the current density requirements of the OLEDs driving proved, to date, to be an insurmountable challenge for a-Si:H TFT technology for use in products, beyond the simple demonstration purposes.

When considering flexible applications, AM-OLED displays on plastic foil have been attracting much attention due to their mechanical robustness, lightweight, flexibility, and thinness. Flexible AM-OLED displays based on organic TFT [15,16], a-Si:H TFT [17], polycrystalline Si TFT [18–20], and metal-oxide TFT [21–25] backplanes have been reported. Particularly, In-Ga-Zn-O-based TFTs show promising results in terms of current driving capacity, uniformity, and stability [26,27].

Most of the In-Ga-Zn-O-based flexible AM-OLED displays are fabricated with high process temperatures typically above 300 °C. This requires thermally stable plastic substrate, for example, polyimide, which is expensive and has lower optical transparency. Alternatives to polyimide are low-cost and transparent polyester films such as polyethylene terephthalate (PET) and polyethylene naphthalate (PEN). Processing on these foils limits maximum processing temperature. A low-temperature process has an added advantage when it comes to the debonding of the fully processed display from the glass carrier. Indeed, the adhesion strength between foil and glass carrier increases with increasing process temperature leading to too strong adhesion between foil and glass that could cause fatal damages such as cracks of dielectric layers, ruptures of electrodes, and delamination of each layer during the debonding process. Therefore, for high-temperature processes, it is necessary to apply complicated and expensive debonding procedures such as laser release [28] that directly affect the manufacturing costs and are to be avoided for production.

In case of PEN foil, maximum processing temperature is limited to 180 °C in order to obtain good overlay accuracy. This prohibits, for instance, the use of typical silicon dioxide (SiO_x) gate dielectric and passivation layers, which are normally fabricated by plasma-enhanced chemical vapor deposition (PECVD) at over 300 °C to obtain low leakage current and good operational TFT stability. Although flexible AM-OLED displays fabricated below 200 °C have been reported [23–25], the TFT properties were poor compared to those processed with temperatures higher than 300 °C. Despite continuous research to make metal-oxide TFT technology compatible with plastic substrates, and although interesting demonstrations have been reported [29], metal-oxide TFT backplanes remain viable, in production environment, only for rigid or conformable, but not truly flexible, AM-OLED displays.

Apart from the technology that can be used in backplanes for driving OLED displays, which is limited by the current density that the TFT is able to pump into the OLED, the other big issue is the driving scheme that has to be used in AM-OLED displays. This aspect has received widespread attention because it determines the complexity of the display, its reliability, and ultimately its manufacturing costs. The basic pixel circuitry used in AM-OLED

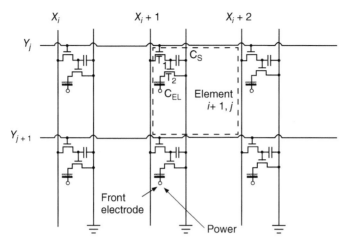

FIGURE 7.1 Original 2T1C circuitry proposed by Brody *et al.* [*Source*: Brody *et al.* [30], Figure 2, p. 741. Adapted with permission of Elsevier.]

displays consists of two transistors and one capacitor where one transistor is used as a switch and the other one as a current source driving the light-emitting diode. This approach was initially proposed by Brody *et al.* for active-matrix EL displays [30] (Figures 7.1 and 7.2).

The amount of current sourced in the OLED is dependent on the parameters of the driving transistor, such as threshold voltage and mobility, rendering this pixel circuitry sensitive to variations of these device parameters. Such variations may be the result of either the TFT fabrication process or the TFT performance degradation during the operational lifetime of the display.

Efforts to overcome the brightness uniformity problems caused by variations in mobility and threshold voltage include external driving architectures [31,32] that regulate the data programmed to the pixel and complex pixel circuits [33–35] that deploy three or more transistors per pixel along with additional control lines.

A typical driving scheme with six TFTs is depicted in Figure 7.3.

A number of simpler circuits, but always with more than two TFTs, have been proposed. Figure 7.4 shows a circuit configuration in which the data line and the sweep line are unified into one data line [36].

Although previous OLED and LCD displays had to have one TFT switch between the data line and the pixel, the proposed circuit configuration needs no such TFT switch. The circuit consists of four TFTs, one capacitor, and five signal lines. Two TFTs in the pixel have been arranged as an inverter circuit, and other two TFTs compose control switches (T4, T5), respectively. TFT switch of T4 is controlled by a reset signal from the reset line to set the

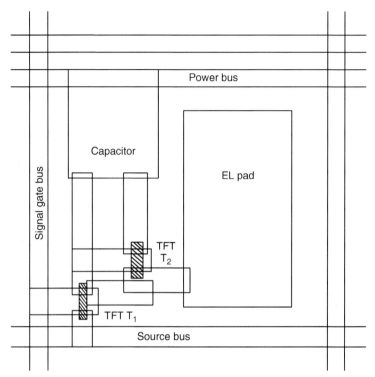

FIGURE 7.2   Schematic of the thin-film version of the original 2T1C circuitry proposed by Brody *et al.* [*Source*: Brody *et al.* [30], Figure 8, p. 744. Adapted with permission of Elsevier.]

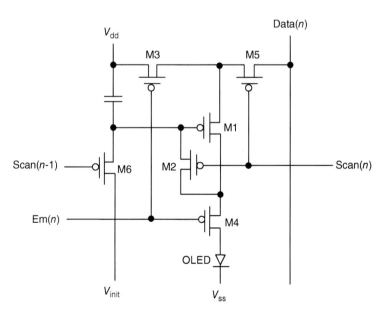

FIGURE 7.3   Circuitry based on a 6T approach to drive an AM-OLED.

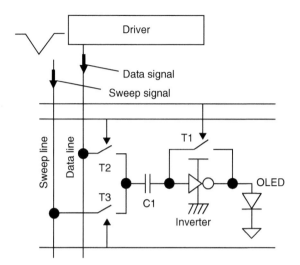

FIGURE 7.4 Circuitry to drive an AM-OLED in which the data line and the sweep line are unified into one single data line. [*Source*: Kageyama *et al*. [36]. Adapted with permission from Wiley.]

inverter to an initial state, and TFT switch of T5 is controlled by an illumination signal from the illumination line to supply source current from the current supply line to the inverter circuit and the OLED element. By designing all components of the pixel circuit with minimum feature of LTPS processing, it would be possible to design pixel configuration with higher aperture ratio than conventional one.

The goal of all of these approaches is to improve brightness uniformity across the display by rendering the drive current independent upon any variations in mobility or threshold voltage. However, these approaches may be difficult to implement in high-resolution mobile displays (cell phones, tablets) as the large area required by the complex pixel circuitry may not be available. Furthermore, the additional components (TFTs and control lines) may result in a lower manufacturing yield, especially for large display sizes.

An approach that utilizes the 2T–1C pixel circuit has been reported by Ono *et al*. [37], but this approach requires a high-speed switching scheme in order to avoid the large detection time constant that is imposed by the storage capacitor tied to the gate of the pixel in conjunction with the capacitance of the OLED.

Alternatively, current feedback programming architectures that can compensate for any variations in threshold voltage or mobility have been proposed.

The basic 2T–1C pixel circuitry within an AM-OLED TFT backplane requires an address (or scan) line, a data line, and a power line. In one

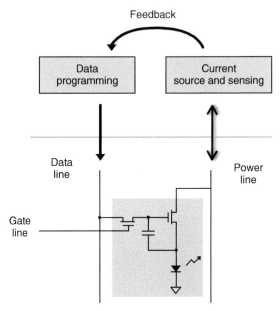

FIGURE 7.5 Schematic of a 2T-1C feedback current programming architecture for driving an AM-OLED display. [*Source*: Charisoulis *et al.* [38]. Adapted with permission from Wiley.]

proposed approach [38], the power line supplies the current that is drawn by the OLED pixels connected to that line and senses the current that is programmed at the addressed pixel. In the driver external to the active-matrix array, a feedback circuit between a data line and a power line is used to adjust precisely the current at the addressed pixel during the pixel addressing time. One approach to simplify the structure (Figure 7.5) is to eliminate the need for an extra sensing line using the power line also for sensing, thus eliminating the need for an extra transistor per pixel as well.

All these efforts, which we briefly recalled, witness how important it is to simplify the driving circuitry and how strategic it is to enable the possibility to use lower quality TFT technology (i.e., TFTs capable of providing lower current density) to fabricate the backplane of the display. As we have extensively discussed in the previous chapters, and as it is visually displayed in Figure 7.6, OLETs combine the function of electrical switching and light emission in a single device.

This basic feature, combined with the fact that the light emission can be modulated by applying a voltage bias to the gate of the OLET, has profound implications both on the driving circuitry design and on the technology that can be used for the backplane in an active-matrix OLET display.

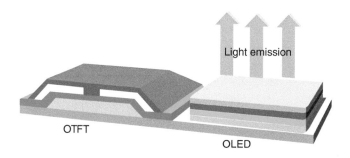

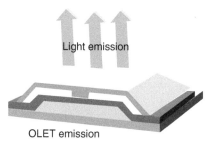

FIGURE 7.6 Schematic rendering of the combination of the switching and emission functionalities in the OLET device.

As far as the driving circuit is concerned, the switching EL OLET device allows the simplification of the basic driving scheme to a 1T approach as shown in Figure 7.7.

Only one switching transistor per EL pixel is used, if the memory effect is guaranteed by the capacitance of the source and drain contacts of the OLET.

The driving circuitry of an active-matrix OLET display may be implemented as in the scheme depicted in Figure 7.8.

The AM-OLET display comprises a driving power supply line ELVDD connected to the first electrode (source) of the OLET, a common power supply line ELVSS connected to the second electrode (drain) of the OLET, a switching element MIM connected to the gate electrode of the OLET, a scan line connected to the switching element, a storage capacitor Cs connected to the switching element and the gate electrode, and a data line connected to the storage capacitor. The switching element MIM is connected to the SCAN line and the gate electrode G of the OLET and turns the gate electrode G on and off in response to a scan signal. The DATA line is connected to the storage capacitor Cs and supplies a data signal to the storage capacitor Cs. Cs is

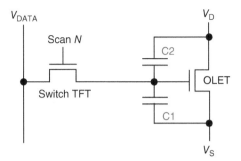

FIGURE 7.7   Basic structure of the active-matrix OLET pixel. Only one transistor is used to switch the OLET, while the capacitance of the source and drain electrodes of the OLET is used to achieve the memory effect of the pixel.

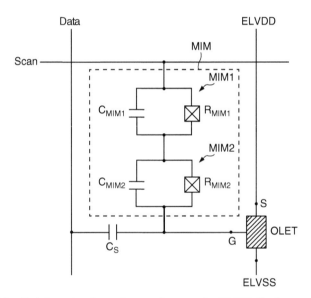

FIGURE 7.8   Driving circuitry of an active-matrix OLET display reported in US 2014/0124759 A1.

connected to the switching element MIM, the DATA line, and the OLET and stores a data signal input to the gate electrode G during one frame.

Multiple OLETs can be integrated as subpixels into a single pixel to design more sophisticated displays as described in US2014/0043307 A1 (Figure 7.9).

Overall, the use of OLETs improves the uniformity of light emission in organic EL displays and simplifies the backplane with respect to AM-OLED displays. In AM-OLED display, the OLED is driven in current mode; thus, a separate device for controlling the OLED is needed. At least two transistors

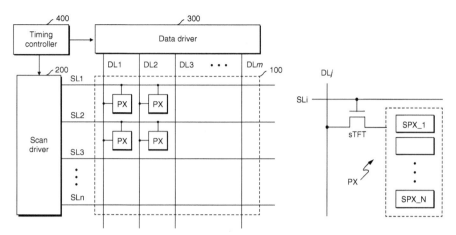

FIGURE 7.9 AM-OLET display comprising N subpixels that form a pixel PX of the OLET display panel as described in US2014/0043307 A1.

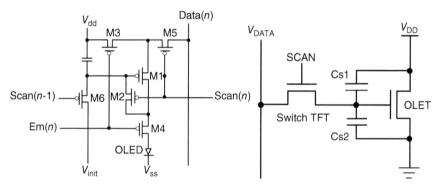

FIGURE 7.10 Direct side-by-side comparison of an AM-OLED display and an AM-OLET display driving scheme.

including a switching transistor for selecting a pixel and a driving transistor for driving an OLED are required. In OLET displays, as in LCD displays, the OLET is voltage-driven, which drastically simplifies the driving circuitry. Figure 7.10 shows a one-to-one comparison of a typical driving scheme of an AM-OLED display and the simplest version of the driving scheme of an AM-OLET display.

The fact the AM-OLET displays are driven as AM-LCD displays enables the use of TFT technologies, which are less performing, and much cheaper, than LTPS. This opens an entire range of possibilities for manufacturing backplanes with generic TFT technologies, including a-Si:H TFTs and OTFTs, whose choice is exclusively determined by the targeted application (rigid, conformable, flexible, or rollable) and not limited by the driving requirements of the display frontplane.

7.2   OLET-BASED SENSING

The concept of chemical sensors involves a change of paradigm in analytical chemistry from general analytical systems to dedicated systems. The chemical information at the basis of the sensing effect is obtained in real time, possibly on site, as a result of the interaction between sensor and chemical(s) in a two-step process: recognition and signal treatment. A change in a physical property as a consequence of a chemical or physical reaction/interaction with a sample occurs or is observed (remote sensing) in the sensor receptor. In cases where the chemical recognition in the receptor does not directly modify an electrical property (resistance, potential), but rather other properties such as heat, mass, or light changes, some type of transducing is required to obtain an electrical signal compatible with the electronic circuits.

Selectivity—the ability to respond primarily to the analyte(s) in the presence of other species—is the key issue with sensors and can be achieved physically, by the selective interaction of the analyte with electrostatic or electromagnetic fields, or chemically, using an equilibrium-based or kinetically based selective interaction with the layer containing the (bio)reagents [39].

The concept of sensors is ubiquitous, but it is most effective when included in portable instrumentation that makes it possible to convert the functions of an entire lab into user-friendly analytical instruments.

The feasibility of optical instruments for sensing depends not only on recognition or transduction principles but also on the entire sensor system, defined by the proper concept of transduction, sensitive layer, data-acquisition electronics, and evaluation software [40]. A careful consideration of sample characteristics (volume, interferents, target concentration, physical properties) and user constraints (type and quality of information needed, assay time, assay frequency, cost, consumables, maintenance, power requirements, technical skill) is required to design handheld/portable instruments [41]. Working under well-defined instrumental and operational conditions for specific analytes and in certain matrices, selectivity and sensitivity are sufficient for direct analysis with less—or even no—prior treatment of the sample.

In this context, lab-on-a-chip (LOC) devices have shown themselves to be highly effective for laboratory-based research, where their superior analytical performance has established them as efficient tools for complex tasks in genetic sequencing, proteomics, and drug discovery applications. However, to date, they have not been well suited to point-of-care or in-the-field applications, where cost and portability are of primary concern. Although the chips themselves are cheap and small, they must generally be used in conjunction with bulky optical detectors, which are needed to identify or quantify the analytes or reagents present. Furthermore, most existing detectors are limited to analysis of a single analyte at a predetermined location on the chip. The lack of an integrated, versatile detection scheme (one that is miniaturized, integrated, wavelength-selective, and able to monitor multiple locations on the chip) is a major obstacle to the deployment of diagnostic devices in the field and has prevented the development of more complex tests where rapid, kinetic, or multipoint analysis is required.

Among the others, optical LOC based on biospecific interaction analysis is of great interest because the technology allows real-time and automated analysis with relatively high capacity.

Fluorescence detection is still the most widely used method for optical sensing systems, due to its superior selectivity and sensitivity. An increasing trend goes toward monolithically integrated sensor systems, merging optical, optoelectronic, and electronic elements (e.g., light sources and/or detectors) into one functional unity fabricated on one common substrate to build LOC systems.

Undoubtedly, light-emitting diodes represent the most commonly used light source in handheld instrumentation. Today, there is a quasicontinuum of commercially available monochromatic LEDs from part of the ultraviolet wavelength (240 nm) to near-infrared (up to 970 nm and discrete wavelengths up to 2680 nm) at several output powers and encapsulations (emphasizing planar and miniaturized surface-mounting devices). The creation of convenient ultrashort-wavelength light emitters that operate in the 200–240 nm range (UV-C) benefits many applications, including air and water purification, germicidal and biomedical instrumentation systems, and ophthalmic surgery tools [42]. Recently, OLEDs have been considered as possible effective light source for sensing. In general, the optoelectronic devices based on organic semiconductors open new possibilities concerning the integration of optoelectronic devices into integrated sensor systems [43]. The ease of processing, based on layer-by-layer vacuum deposition, and the possibility to deposit, as an example, organic photodiodes (OPDs) practically on any substrates[44], as well as the possibility to tune their spectral response [45] make organic electronics attractive for integrated systems, especially allowing facile integration with planar chip-based systems. Until now, different integrated optical sensor systems utilizing organic electronics have been described: some of these concepts are based on integrated OLEDs [1,46] as light sources, others use OPDs as integrated optical detectors for chemiluminescence [47,48] and fluorescence [49,50] (Figure 7.11).

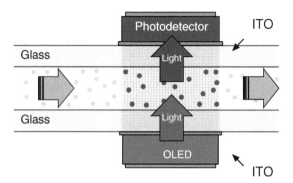

FIGURE 7.11 Simple schematic of a polymer detection system developed at Molecular Vision (https://www.abingdonhealth.com/molecular-vision/).

In LOC sensors, the organic diode architecture under electrical excitation can be implemented as light sources or as photodetectors in reverse bias by illuminating the active layer so as to generate a measurable electrical current [51]. Both devices rely on the use of transparent electrodes (typically indium tin oxide—ITO) that define the emission/illumination side of the device and need to be fabricated on the glass surface of the microfluidic chip. Indeed, the integrated detectors utilize combinations of OLEDs and organic photodiodes arranged on the top and bottom surfaces of the (transparent) microfluidic chips.

Labeled analytes passing through the detection volume absorb photons from the OLED and may subsequently re-emit photons, which are detected by the photodetector. Owing to the layer-by-layer deposition processes for the OLEDs and organic photodiodes and the planar structure of typical microchip devices, the photonic components may be integrated into existing chip structures.

Importantly, since they may be fabricated by print-deposition procedures, parallel arrays of light sources and photodetectors can be fabricated with ease, allowing for the straightforward detection of multiple analytes via both end-point and kinetic assays. In short, the use of OLEDs and organic photodiodes allows for the creation of devices that are much smaller, simpler in structure, easier to manufacture, and significantly more versatile than the off-chip laboratory-scale optics currently used for detection.

A remarkable integrated sensor chip was recently reported consisting in fluorescent sensor spots surrounded by integrated ring-shaped organic photodiodes where the fluorescence emission of the individual spots was detected [52]. The monolithic integration of sensor spot and detector on one common substrate as well as the special ring shaped form of the photodiodes guarantees that a maximum of the fluorescent light emitted from the sensor spots is

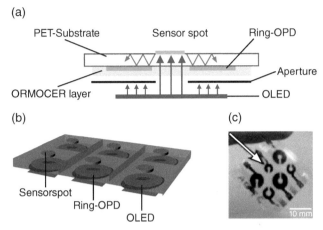

FIGURE 7.12 (a) Schematic setup of the sensor, consisting basically of fluorescent sensor spots on top and ring-shaped photodiodes on the backside of the substrate. The sensor spots are excited by an OLED positioned below an aperture. (b) 3D illustration of a multisensor chip. (c) Photographic image of the sensor chip with six integrated OPDs and sensor spots in different sizes. [*Source*: Lamprecht *et al.* [52]. Adapted with permission from Wiley.]

collected. Filterless discrimination between excitation light and generated fluorescence light is an essential advantage of the presented design (Figure 7.12). This simplifies the detection system by minimizing the number of required optical components and offers a means of creating compact and potentially sensitive integrated sensor devices.

The feasibility of the integrated sensing concept was demonstrated by realizing an integrated oxygen sensor device. The sensing scheme is based on dynamic quenching of the photoluminescence of an immobilized indicator dye by molecular oxygen. A method called light harvesting is applied to enhance the brightness of the sensing spots. Antenna dyes collect the excitation light of the OLED and transfer the energy to an oxygen-sensitive acceptor dye. The authors reported that this method yielded up to fivefold higher signals compared to a direct excitation of the indicator [53].

Another promising approach in realizing sensor device for monitoring biomolecular activities is based on the surface plasmon resonance (SPR) phenomenon given its advantages of real-time monitoring, label-free assay capability, and high sensitivity. Traditionally, the SPR configuration based on the Kretschmann [54] prism coupler has been the most popular SPR sensor approach because the SPR phenomenon in this configuration is simple to establish and analyze. The principle of the SPR sensor is based on exciting the surface plasmon wave in the thin metal–analyte medium interface using p-polarized light at a specific angle through a high-refractive-index prism. Subsequently, the reflectivity dip profile is obtained, and the reflectivity dip shifts once the refractive index of the medium is changed [55]. The common commercial SPR sensor system is large in size and weight, requires high power consumption, and involves high cost [56–60]. Hence, the challenges in SPR sensor development are miniaturization, cost-effectiveness, and portability of the sensor structure [61]. Indeed, the design of an SPR sensor that uses an OLED light source attached to the prism was reported and it not only enhanced the system miniaturization and simplified the assembly procedures but also attained a limit of detection of 6×10^{-4} RIU (refractive index unit)[62]. Nevertheless, there are still many obstacles that prevent the implementation of this design. First, compared with a laser, an OLED emits light over a wide angle. Thus, determining how to reduce the output angle of the light to be focused in the direction perpendicular to the face of the prism is important to enhance the surface plasmon wave. Second, a complicated configuration of the polarizer is required to obtain pure p-polarized light incident on the SPR system. A conventional polarizer, such as a polarizing beam splitter (PBS), presents large cubic size ($\sim 1 \text{ cm}^3$) compared to the OLED thickness ($\sim 2 \text{ mm}$); therefore, on an SPR system using an OLED, the polarizer is used on the path along which the light is reflected [62]. Using this polarizer configuration, the sensing metal layer is coupled by nonpolarized light rather than by pure p-polarized light. A recent study indicated that SPR can be enhanced by the use of both s-polarized and p-polarized light via the insertion of an anisotropic thin film below the metal [63]. However, for an SPR sensor design using a conventional noble metal film, such as Au or Ag, only p-polarized light is able to enhance the SPR [54,61,64].

An SPR sensor based on an OLED using a brightness enhancement film (BEF) together with a giant birefringent optical (GBO) film in order to provide the correct polarizer configuration for an SPR system, without any significant change in the light source size (Figure 7.13), was recently demonstrated [65]. The BEF is integrated with the OLED to improve the light intensity and allow non-Lambertian angular emission [66,67]. Furthermore, GBO film is used as a reflective polarizer [68] next to the BEF film to obtain p-polarized light and couple the surface plasmon wave. The integrated light source combined with an OLED-BEF-GBO film enables cost-effectiveness and a simple optical construction while providing optical enhancement of the SPR system.

As it was demonstrated by the previously reported examples, the implementation of vertical-architecture organic devices (as OLEDs and OPDs) in LOC sensors offers a powerful platform for biochemical analysis, enabling

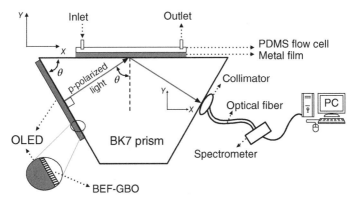

FIGURE 7.13 Top view of the integrated SPR sensor system (not to scale) using an OLED light source covered by BEF and GBO film. [*Source*: Prabowo *et al.* [65], Figure 1, p. 426. Adapted with permission of Elsevier.]

detection down to the picomolar level—more than sufficient for the majority of biochemical applications [69]. However, to achieve such sensitivities, relatively sophisticated support instrumentation is needed to firstly amplify, then condition, and finally quantify the miniscule signals that are obtained when analyzing dilute solutions at the microscale. In addition, although far easier to fabricate and integrate than their inorganic counterparts, the (multi-layered) organic LEDs and photodiodes nonetheless require the use of an ITO coating, which is known to be a critical issue of the OLED/OPD technology, and multiple processing steps that add complexity, limit production yields, and ultimately increase costs.

In this scenario, the use of intense nanoscale light sources with controlled light-emission pattern and tunable photonic integration in optoelectronic systems is a key enabling step for the development of effective, low-cost, compact, portable, and highly sensitive sensors.

As we have seen throughout the book, one of the attractive characteristics of OLETs is the possibility to vary the spatial position of the light emission zone across the transistor channel by acting on the gate voltage. This feature, which is unique to the planar field-effect architecture, may be exploited for defining innovative excitation and detection scheme in next-generation optical sensors by fine-tuning the optical pig-tailing within the photonic components in the final LOC.

Moreover, the in-plane light generation distant from charge-injecting electrodes together with the stripe-like and micrometer-wide emission area is a distinguished photonic feature intrinsic to OLETs, which may allow the following: (i) the prevention of the implementation of optics in light coupling, (ii) higher integration among photonic components by monolithic device fabrication onto functional layers (i.e., charge-injecting and gate electrodes), and (iii) the optimization/simplification of the coupling with the micrcofluidic chip given that the typical length of the OLET channel (30–100 μm) is well suited for alignment with fluidic microchannels.

Moreover, it is expected to push down the fabrication costs by developing a newer generation of diagnostic devices, in which the OLEDs and organic photodiodes are replaced by light-emitting and light-sensitive organic field-effect transistors. Different attempts in validating the photonic field-effect technological platform for sensing have been proposed in the past years, given the recent improvements in performance capabilities of organic light-emitting and light-sensing field-effect transistors realized with low-cost techniques. For instance, Figure 7.14 reports the concept of a possible integration scheme of the photonic field-effect architecture into a microfluidic channel, which was developed within the EU Project PHOTO-FET.

The use of transistors in place of diodes has four principle benefits. Firstly, they can be fabricated in a top gate geometry wherein the source and drain electrodes are predeposited on the substrate (in this case, the microfluidic chip), ensuring optimal registration between the organic components and the underlying fluidic architecture. Secondly, the devices can be completed by depositing a single layer of organic semiconductor followed by the gate dielectric/electrode, thus avoiding the need for complex multilayered structures and the use of ITO. Thirdly, in-plane light generation by the transistors, which is directly generated at the interface with the microfluidic chip surface, provides improved optical coupling into waveguiding structures, leading to significant performance gains.

Fourthly, current-multiplying auxiliary transistors can also be included, as either drivers for the light-emitting transistors or amplifiers for the phototransistors. The use of such auxiliary transistors allows for simpler support instrumentation and enhances the signal-to-noise characteristics by enabling immediate amplification of the signal at the point-of-generation, thus avoiding the inadvertent amplification of interferences that would otherwise occur if the amplifier were located at a distance from the photodetector.

However, OLETs that are compatible for integration into optomicrofluidic sensors for fluorescence detection have to show specific optoelectronic characteristics: (i) external EL quantum efficiencies in the range of 1–10%; (ii) brightness in the range of 500 cd/m^2; (iii) high ambipolar carrier mobilities with values of $\mu_e = \mu_h \sim 0.1$–$1\,cm^2/V\,s$; (iv) operating speeds in the range 10–100 kHz; (v) tunable wavelength emission over a large spectral range in the case of photoluminescence-based optical sensors; and (vi) compatibility with the technologies subsequently used for the fabrication of the fully integrated photonic biosensing component.

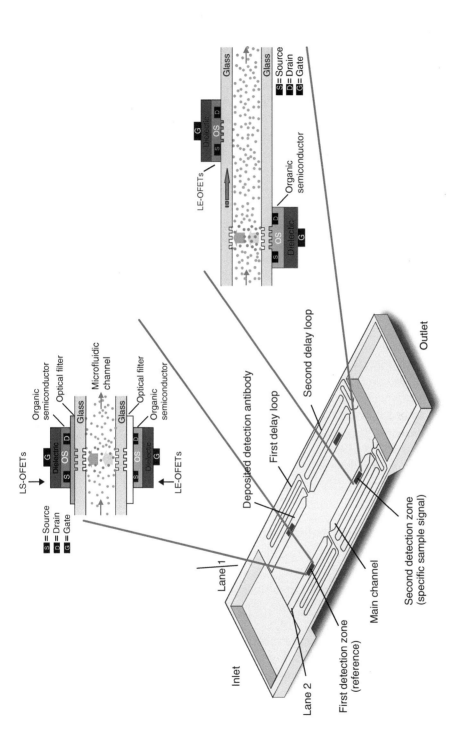

FIGURE 7.14 Generic scheme of organic photonic field-effect device integration into biosensing microfluidic chips. [Courtesy of PHOTO-FET (EU project FP7-ICT-248052).]

Once the target performances are achieved, it is expected that the introduction in the point-of-care diagnostic market of sensors based on the organic photonic field-effect platform will allow the realization of portable diagnostic platforms with multiple, quantitative, and highly sensitive detection methods, which are capable of operating without the requirement for a dedicated desktop reader.

Broadening the application area to the realization of advanced biomedical tools for biodiagnostics and therapy in living systems, highly efficient/ effective organic optoelectronic devices may also provide optoelectronic stimulation, manipulation, and read-out of bioelectrical activity of excitable cells such as neurons and cardiomyocytes *in vitro* and *in vivo*.

Nowadays, the electronic devices interfacing with living systems have become a stringent necessity in clinical practice to improve the capability of diagnosis and treatments for a broad spectrum of diseases. Devices such as cardiac pacemakers and cochlear implants monitor, stimulate, and modulate electrically active cells, restoring lost function, and improving the quality of life of patients affected by cardiovascular pathophysiologies as well as hearing loss [70].

Particularly, a significant effort is needed to provide novel tools for neuroscience electrophysiological investigations and for therapeutic applications that enable real-time recording and manipulation of dynamic communication processes between neural cells [71].

One of the most striking properties of neurons is their ability to form complex circuits devoted to the processing and propagation of information. Understanding how neurons interact within networks, how they process information, and how they adapt to novel conditions is one of the most interesting questions in neuroscience.

In recent years, the use of inorganic field-effect transistors for live cell recordings has been well documented *in vitro* [72,73], even though characterized by low signal-to-noise ratio. Recently, recordings from optically induced epileptic seizures have been attained using surface electrode arrays based on thin-film silicon technology [74]. However, inorganic semiconductors are not ideally suited for interfacing to living systems because of their time-limited biocompatibility, mechanical stiffness, and rigid form factor. These data suggest that the future of *in vivo* recording should be based on innovative flexible materials that integrate with the surrounding environment without detrimental impact on tissue physiology.

In this scenario, organic semiconductors (molecular systems and polymers) are promising biointerfacing materials [75] due to their chemical similarities to biological systems (carbon-, oxygen-, nitrogen-based composition), thus offering the prospect of improved compatibility in devices intended for neurophysiological and pathophysiological investigations and applications. Softness and flexibility of organic materials, as compared to traditional

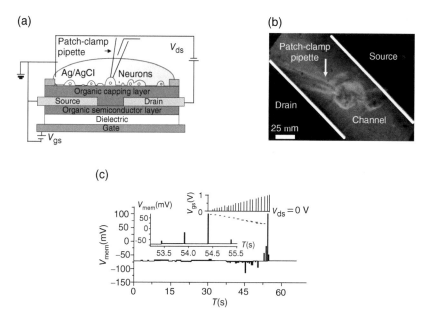

FIGURE 7.15 (a) Schematic of the O-CST architecture and of the experiment for extracelluar/intracellular stimulation and intracellular/extracelluar recording. (b) Photograph of the neuronal cell patched in the channel of the transistor. (c) Current-clamp trace recorded in the patch-clamp experiment before and upon the reported pulsed O-CST stimulation protocol (upper inset). Dashed line indicates a zoom of the trace portion, which highlights the action potential profile. [*Source*: Benfenati *et al.* [76]. Adapted with permission of Nature Publishing Group.]

silicon-based electronic materials, offer improved compatibility with the neural tissue, whereas the combined ionic and electronic conductivity enables the use of organic electronic devices as direct signal transductors of the neural systems [75].

Recently, a silicon-free transparent FET based on N,N'-ditridecyl-perylene-3,4,9,10-tetracarboxylic diimide (P13) was reported [76], which provides bidirectional stimulation and recording of the dorsal root ganglion (DRG) primary neurons (Figure 7.15a). The device, which is named organic cell stimulating and sensing transistor (O-CST), is compatible with the growth, differentiation, and functionality of DRG neurons, provides their electrical stimulation, and allows both extracellular current recording and optical read-out of membrane potential modulation.

Particularly, it was defined as a device characterization protocol by cross-correlating frequency-modulated field-effect transistor characteristics with well-assessed electrophysiological methods, so that neuron stimulation by the O-CST and simultaneous recording of the intracellular membrane potential modulation by whole-cell patch-clamp in current-clamp mode are correlated (Figure 7.15b).

Once a train of increasing voltage pulses was applied to the gate electrode (Figure 7.15c, inset: from $V_{gs} = 0$ V to $V_{gs} = 1$ V with linear increasing steps of 20 mV, pulse duration 200 µs, frequency rate 2 Hz), V_{mem} measured intracellularly first responded with short negative transients at the beginning of the stimulation ramp. Positive depolarizing spikes resembling action potentials (Figure 7.15c, zoom) were recorded next when the V_{gs} bias reached values around 800 mV, thus demonstrating the modulation of the cell electrophysiological activity induced by the biased organic device.

In order to fully validate the O-CST device platform, it was mandatory to prove the bidirectional communication between the device and the neurons by demonstrating the stimulation and recording of neuronal electrical firing. Moreover, the O-CST detection sensitivity had to be compared with the benchmark and well-consolidated method for extracellular recording, which is the MEA (multielectrode array) approach [77].

Indeed, it was demonstrated that the O-CST device has a 16 times higher signal-to-noise ratio than the MEA, both coupled to the same neuronal preparation and to well-assessed cortical cells with efficient MEA coupling. This unprecedented result highlights that the development of organic devices based on field-effect architecture specifically engineered for single neuron interfacing, as well as of arrays for manipulation of mature neural networks, might enable investigations of the nervous system and broaden the scope for highly efficient *in vitro* drug screening at low analyte concentrations.

It is self-evident that the introduction of further functionalities, such as light emission capability, in the working principle of such field-effect transistor-based organic devices, will allow the realization of innovative cell stimulation tools with enhanced selectivity and efficiency. Indeed, the optical stimulation provides new opportunities, enhancing temporal and spatial resolution with respect to the more traditional electrical methods.

The recent development of optoelectronics and neural photosensitization tools, such as channelrhodopsin-2 (ChR2)10, voltage-sensitive dyes, and fusion proteins, offers new opportunities for noninvasive, cell-specific neuronal optical stimulation [78,79]. Previously, spatial and temporal control of photosensitized neurons has been limited by the lack of appropriate optoelectronic devices that could provide two-dimensional stimulation with biocompatibility, sufficient irradiance [71], appropriate color tunability, controlled-size active area, mechanical flexibility, and low-cost processing [80].

In this scenario, OLET technology can play a prominent role in defining disruptive methods for investigating neuronal circuits and their functions, given the inherent photonic features of this light source, such as the spatially controlled, voltage-tunable, and micrometer-wide emission area, the planar uncovered geometry that allows intimate interaction with living systems, the high light-emission efficiency and brightness.

7.3 OPEN ISSUES AND NEXT DEVELOPMENT TARGETS

After its invention in 2003 (EP 1609195B9; US 8497501B2) by Muccini *et al.*, OLETs experienced a great development, particularly in the last few years. Indeed, by taking advantage of the materials and structural solutions developed for OLEDs and OTFTs, the OLET technology has been developed to a level that makes it possible its application in manufacturing environment. However, it is clear that issues remain that need to be thoroughly addressed to enable the full deployment of the potential of the OLET technology and its penetration in the display, health-care, and sensor markets. The key aspects that deserve attention and that are common needs for all possible applications concern color gamut, power efficiency, lifetime, and reliability.

7.3.1 Color Gamut

OLETs based on different classes of materials and with different structures of the active region have been demonstrated to be able to emit light of different colors. The fundamental red, green, and blue colors have been reported by many research groups using a variety of emitters, which demonstrates that it is possible to develop OLETs with specific color coordinates and combine them in multicolor integrated systems. However, building on the experience of the OLED technology, it is to be expected that some issues will have to be specifically addressed, particularly concerning the quality of the color gamut and the efficiency and lifetime of the blue emission.

The need for tailored emission emerges for applications where not only the color but also the specific emission spectrum is important to match the spectral requirements of the illumination source. This might be the case for OLETs integrated into systems for health care, optical sensing, on-chip spectroscopy, and data communications.

In the case of display technology, important are the color gamut and the ability to render vivid and natural colors. Indeed, for this application, each color available in the display is obtained by the combination of the fundamental red, green, and blue colors (Figure 7.16).

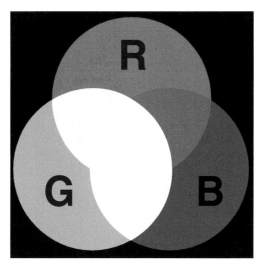

FIGURE 7.16 Additive color mixing: fundamental concept showing three overlapping light bulbs in vacuum, adding together to create white.

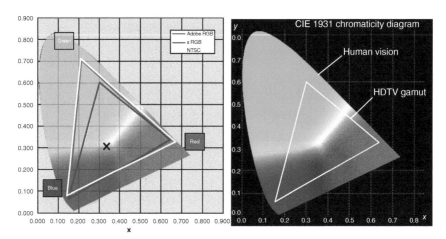

FIGURE 7.17 CIE chromaticity diagrams showing the human vision, the standards NTSC, sRGB, Adobe RGB, and representative HDTV Gamut.

The color scale of the display and, ultimately, its color gamut are determined by the combination of the available fundamental RGB colors (Figure 7.17).

For multicolor displays, it is therefore crucial to achieve color coordinates for the red, green, and blue that enable a color vision experience similar to that offered by the technologies available in the market. Similarly to the issues faced by the OLED technology, it is likely that the major challenge for OLETs will be the development of a deep, efficient, and stable blue. Indeed, blue emission from organic materials undergoes fundamental energetic and electronic constrains, which affect its emission spectrum, efficiency, and molecular structural stability under operational conditions. The search for a satisfactory solution that combines at one time all of these properties must necessarily take into account a trade-off between photonic characteristics, operational stability, and pixel resolution. The fundamental differences in the device architecture and device physics of OLETs with respect to OLEDs pose different constrains to these problems and are likely to offer new solutions to the challenge of achieving satisfactory blue EL emission from organic materials.

7.3.2 Power Efficiency

As we have discussed in Section 7.1, the great interest attracted by AM-OLED displays is due to the combination of its distinguishing advantages such as lightweight, short response time, low driving voltage, wide color gamut, wide viewing angle, high brightness, and low power consumption. Particularly important for mobile applications is the power consumption, whose value makes a specific display technology suitable or not for integration into appliances with different uses (e-readers, wearable devices, smartphones, notebooks, etc.). The power efficiency is also extremely important in nondisplay applications, when the OLET is used as onboard light source in portable analytic and sensing sets. The efficiency of the OLET device depends on both the material efficiency and the device structure [81]. Key factors that determine the AM-OLET display power consumption are, in addition to the OLET device efficiency, line resistances on the substrate and drive architecture. The first key requirements for low-power-consumption displays are, therefore, efficient emitting materials, low operating voltages, and high light extraction efficiency.

As we have seen in Chapter 6, light emission efficiency of OLETs can be expressed by the following basic relation:

$$\eta_{ext} = \gamma \, \Phi_{PL} \chi \sigma_{out}$$

where η_{ext} is the external quantum efficiency of OLETs; σ_{out} is the outcoupling efficiency of the generated light; χ is the fraction of total exitons formed, which result in radiative transition; γ is the fraction of opposite charges, which combine to form the exiton; Φ_{PL} is the intrinsic quantum efficiency for radiative decay (see also Section 6.1 and Equation 6.1 where these processes are also discussed).

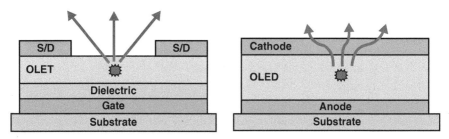

FIGURE 7.18   Schematics of the light extraction paths in the OLET and the OLED architecture.

Clearly, phosphorescent EL materials are very effective in increasing the light emission efficiency of OLETs because they use the triplet state of an organic molecule for light emission. Phosphorescence is distinguished from fluorescence in that fluorescence emits light from only singlet state, while phosphorescence uses triplet states for light emission and can in principle take advantage of the singlet state as well. Therefore, theoretical maximum internal quantum efficiency for phosphorescence reaches 100% compared with 25% for fluorescence. Indeed, it is known that the relative contribution of fluorescent blue color for the power consumption in phosphorescent AM-OLED is typically of the order of 50% of the total device power consumption because of the low efficiency of the blue emitter. Therefore, the development of blue phosphorescent OLET is highly desirable.

The light extraction can also be favored in the OLET architecture compared to the OLED one. Indeed, OLET can take advantage of the planar arrangement of the injecting electrodes to avoid cavity effects and light losses at interfaces (Figure 7.18).

Also, the viewing angle is expected to be wider in OLETs, since the natural isotropic emission occurring in the emitting layer is much less distorted before it reaches the open space, than it is in the sandwich structure of the OLED.

It is therefore clear that, although the implementation of low-power-consumption OLET devices in AM-OLET displays will be a manufacturing challenge, this technology has great potential also in relation to this key requirement.

7.3.3   Lifetime

The lifetime and reliability characteristics are key figures for any light-emitting technology and are even more critical for organic-based technologies, which are known to be highly sensitive to oxygen and moisture and subject to severe

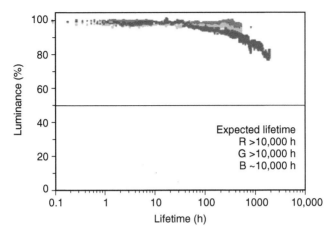

FIGURE 7.19 Lifetime of RGB colors measured in an AM-OLED display at a white brightness of $100\,cd/m^2$. [*Source*: Lee *et al.* [82], Figure 4, p. 147. Adapted with permission of Elsevier.]

irreversible degradation. Representative lifetime curves of RGB colors in an AM-OLED display are plotted in Figure 7.19 [82].

The lifetime of RGB was measured at white $100\,cd/m^2$ condition, which corresponds to 600, 800, and $400\,cd/m^2$ for RGB, respectively. In an OLED, the lifetime of phosphorescent materials is greatly dependent on the hole and electron charge balance and the distribution of the recombination region in the light emitting layer. Therefore, the control of the thickness of charge transport layers is critical to the lifetime of the phosphorescent materials. Particularly, the thickness of electron transport layer and hole blocking layer is important for the lifetime and efficiency. Thick electron transport layer and hole blocking layer lead to the increase of luminance efficiency but poor lifetime. The optimum thickness for the lifetime does not correspond to the optimum structure for the luminance and efficiency.

These observations, which hold strictly true for the OLED stack and architecture, provide important hints to appreciate the potential that the OLET technology has in terms of lifetime and device reliability. The planar device structure and the lateral charge transport, coupled to the third control electrode that allows to balance the electron and hole currents in the device active region, as well as the tuning of the position of the recombination area, are key factors to better optimize the fundamental electronic and excitonic processes, which ultimately lead to lower degradation and longer lifetime. It is worth highlighting that, while the control of the fundamental excitonic processes in OLEDs is demanded to the engineering of the OLED stack, in

OLETs, the electric field and charge distribution is determined by a combined effect of the device structure and the applied biases. This feature provides a much higher degree of freedom to drive the OLET at the most appropriate bias point to prompt the device lifetime and balance aging effects and instabilities.

In addition to intrinsic degradation, which is correlated to the device architecture and charge distribution in the organic layers, organic electronic devices are extremely sensitive to the environment and require high performance protection. It is known that the interaction of an OLED with the ambient atmosphere results in the formation of black spots in electroluminescence [83–85]. These spots are a consequence of oxidation of the cathode through pinholes in the cathode itself. These pinholes are mainly induced by particles that are present during the processing of the device, and encapsulation is a prerequisite to prevent the black spot problem. OLEDs are typically encapsulated or sealed in a metal or glass coverlid, and the cavity between the device and the coverlid accommodates a getter that traps and consumes any water penetrating through the edge seal. This gives excellent environmental protection at the expenses of mechanical flexibility and costs. Indeed, an effective encapsulation can represent up to 50% of the cost of the finished system [85].

For large-area or flexible displays, this solution is simply not viable for reasons of costs and functionality. The availability of a high-performance thin-film barrier is the most critical challenge in upscaling and commercializing flexible displays, and this challenge holds for any organic-based technology (Figure 7.20).

The goal to develop a flexible, low-cost, thin-film-based barrier technology, which offers sufficient environmental protection to fulfill the lifetime specifications of the display products and facilitate the upscaling to a manufacturable level, is indeed a game changer. Conventionally, the quality of thin-film barriers for OLED is tested by means of a calcium (Ca) test, which results in a water vapor transmission rate (WVTR) [84]. For proper encapsulation of OLEDs, it is considered that the WVTR should be lower than 10^{-6} g H_2O/m^2 day. However, in addition to the calcium test, the occurrence and growth rate of black spots should also be considered, and this is probably a much more reliable criterion to evaluate the effectiveness of a barrier for a specific type of EL device. It is clear that the device structure and the different positioning of the constituting building blocks in the OLET may once more offer a different platform to protect selectively the most delicate and exposed parts of the device and develop a tailored protection paradigm for OLETs. However, it remains that the availability of a high-performance thin-film flexible barrier is a key enabler for the realization of flexible OLET products.

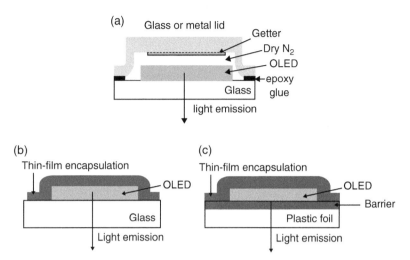

FIGURE 7.20 Schematic representation of (a) standard rigid encapsulation, (b) thin-film encapsulation of an OLED on glass, and (c) encapsulation of a flexible OLED using a bottom barrier and a top thin-film encapsulation.

7.4 CONCLUSIONS

It is clear from all the discussions that the key fundamental features of OLETs make them candidates to impact a number of application fields. It is worth noting that, beyond the considerations on the status of the current developments stage, the potential of OLETs stems from fundamental architectural features, which will deploy their power as the technology advances along its development path. The application that might benefit, first, from the use of the OLET technology is the flat-panel display technology. Indeed, the possibility to combine an OLET frontplane with any available backplane technologies, including low-cost a:Si-TFT and rollable OTFT, is a game changer for the field of active-matrix displays. The timing for deploying the potential of this technology will of course depend on the efforts that the international industrial community will devote to focused development programs.

REFERENCES

[1] J. Shinar and R. Shinar, *J. Phys. D* **2008**, *41*, 133001.
[2] J. W. Parks, M. A. Olson, J. Kim, D. Ozcelik, H. Cai, R. Carrion Jr., J. L. Patterson, R. A. Mathies, A. R. Hawkins, and H. Schmidt, *Biomicrofluidics* **2014**, *8*, 054111.

[3] E. B. Namdas, I. D. W. Samuel, D. Shukla, D. M. Meyers, Y. M. Sun, B. B. Y. Hsu, D. Moses, and A. J. Heeger, *Appl. Phys. Lett.* **2010**, *96*, 043304.

[4] C. Koos, *et al.*, *Nat. Photon.* **2009**, *3*, 216.

[5] C. Sciascia, N. Martino, T. Schuettfort, B. Watts, G. Grancini, M. R. Antognazza, M. Zavelani Rossi, C. R. McNeill, and M. Caironi, *Adv. Mater.* **2011**, *23*, 5086.

[6] T. Manaka, S. Kawashima, and M. Iwamoto, *Appl. Phys.* **2010**, *97*, 113302.

[7] M. Ullah, K. Tandy, J. Li, Z. Shi, P. L. Burn, P. Meredith, and E. B. Namdas, *ACS Photonics* **2014**, *1*, 954.

[8] S. Lee, A. Badano, and J. Kanicki, *Proc. SPIE 4800* **2002**, 156.

[9] G. E. Jabbour, S. E. Shaheen, M. M. Morrell, B. Kippelen, N. R. Armstrong, and N. Peyghambarian, *Opt. Photon. News* **1999**, *10*(4), 24.

[10] J. W. Allen, *J. Lumin.* **1994**, *60*, 912.

[11] D. E. Mentley, *Proc. IEEE* **2002**, *90*, 453.

[12] Special issue on small molecule and polymer organic devices, *IEEE Trans. Electron Device*, August **1997**.

[13] H. K. Chung and K. Y. Lee, *SID'05 Tech Digest* **2005**, 956.

[14] S. Terada, G. Izumi, Y. Sato, M. Takahashi, M. Tada, K. Kawase, K. Shimotoku, H. Tamashiro, N. Ozawa, T. Shibasaki, C. Sato, T. Nakadaira, Y. Iwase, T. Sasaoka, and T. Urabe, *SID Symposium Digest* **2003**, *34*, 1463.

[15] M. Suzuki, *et al.*, Proc. IDW'08 **2008**, 1515.

[16] M. Katsuhara, *et al.*, *SID Symposium Digest* **2011**, *9*, 656.

[17] R. Ma, *et al.*, *J. Soc. Info. Display 18*(1), **2010**, 50.

[18] S. An *et al.*, *SID Symposium Digest* **2010**, *41*, 706.

[19] S. An, J. Lee, Y. Kim, T. Kim, D. Jin, H. Min, H. Chung, and S. Kim. *SID Symposium Digest* **2011**, *42*, 194.

[20] H. H. Hsieh, *et al.*, *Proc. IDW'12* **2012**, 1859.

[21] M. Noda, *et al.*, *SID Symposium Digest* **2012**, *74*, 998.

[22] H. Yamaguchi, *et al.*, *SID Symposium Digest* **2012**, *74*, 1002.

[23] Y. Nakajima, *et al.*, *SID Symposium Digest* **2012**, *74*, 271.

[24] K. Miura, *et al.*, *SID Symposium Digest* **2011**, *42*, 21–24.

[25] A. K. Tripathi, *et al.*, *Proc. IDW'12* **2012**, 763.

[26] K. Nomura, H. Ohta, A. Takagi, T. Kamiya, M. Hirano, and H. Hosono, *Nature* **2004**, *432*, 488.

[27] T. Kamiya, *et al.*, *Sci. Technol. Adv. Mater.* **2010**, *11*, 044305.

[28] I. French, *et al.*, *SID Symposium Digest*, 2007, *38*, **1680**.

[29] Y. Fukui, *et al.*, *SID Symposium Digest 18*(1), **2013**, 203.

[30] T. Brody, F. C. Luo, Z. P. Szepesi, and D. H. Davies, *IEEE Trans. Electron Dev.* **1975**, *22*, 739.

[31] H. J. In and O. K. Kwon, *Jpn. J. Appl. Phys.* **2010**, *49*, 03CD04.

[32] H. J. In, K. H. Oh, I. Lee, D. H. Ryu, S. M. Choi, K. N. Kim, H. D. Kim, and O. K. Kwon, *IEEE Trans. Electron Devices* **2010**, *57*, 3012.

[33] S. J. Ashtiani, G. R. Chaji, and A. Nathan, *J. Display Technol.* **2007**, *3*, 36.

[34] C. L. Lin, T. T. Tsai, and Y. C. Chen, *J. Display Technol.* **2008**, *4*, 54.

[35] N. Fruehauf, P. Schalberger, M. Herrmann, and A. Vielwock, Active-Matrix Flatpanel Displays and Devices (AM-FPD), 19th International Workshop **2012**, 277.

[36] H. Kageyama *et al.*, *SID Symposium Digest* **2003**, *34*, 96.

[37] S. Ono, K. Miwa, Y. Maekawa, and T. Tsu-Jimura, *IEEE Trans. on Electron Devices* **2007**, *54*, 462.

[38] T. Charisoulis, M. Troccoli, D. Frey, M. K. Hatalis, *SID Symposium Digest* **2013**, *44*, 465.

[39] J. Janata, *Principles of Chemical Sensors*, 2nd edn, Springer, **2009**.

[40] G. Gauglitz, *Anal. Bioanal. Chem.*, **2005**, *381*, 141.

[41] F. S. Ligler, *Anal. Chem.*, **2009**, *81*, 519.

[42] A. Khan, K. Balakrishnan, and T. Katona, *Nat. Photon.* **2008**, *2*, 77.

[43] R. Pieler, E. Fusreder, and M. Sonnleitner, *Proc. SPIE* **2007**, *6739*, 673919.

[44] B. Lamprecht, R. Thunaer, M. Ostermann, G. Jakopic, and G. Leising, *Phys. Status Solidi A* **2005**, *202*, R50.

[45] B. Lamprecht, R. Thunauer, S. Kostler, G. Jakopic, G. Leising, and J. R. Krenn, *Phys. Status Solidi RRL* **2008**, *2*, 178.

[46] V. Savvateev, Z. Chen-Esterlit, J. W. Aylott, B. Choudhury, C.-H. Kim, L. Zou, J. H. Friedl, R. Shinar, J. Shinar, and R. Kopelman, *Appl. Phys. Lett.* **2002**, *81*, 4652.

[47] O. Hofmann, P. Miller, P. Sullivan, S. T. Jones, J. C. deMello, D. D. C. Bradley, and A. J. de Mello, *Sens. Actuators B*, **2005**, *106*, 878.

[48] O. Hofmann, X. Wang, A. Cornwell, S. Beecher, A. Raja, D. D. C. Bradley, A. J. deMello, and J. C. deMello, *Lab Chip* **2006**, *6*, 981.

[49] E. Kraker, A. Haase, B. Lamprecht, G. Jakopic, C. Konrad, and S. Kostler, *Appl. Phys. Lett.* **2008**, *92*, 033302.

[50] A. Pais, A. Banerjee, D. Klotzkin, and I. Papautsky, *Lab Chip* **2008**, *8*, 794.

[51] J. L. Segura, N. Martin, and D. M. Guldi, *Chem. Soc. Rev.* **2005**, *34*, 31.

[52] B. Lamprecht, T. Abel, E. Kraker, A. Haase, C. Konrad, M. Tscherner, S. Köstler, H. Ditlbacher, and T. Mayr, *Phys. Status Solidi RRL* **2010**, *4*, 157.

[53] T. Mayr, S. M. Borisov, T. Abel, B. Enko, K. Waich, G. Mistlberger, and I. Klimant, *Anal. Chem.* **2009**, *81*, 6541.

[54] E. Kretschmann, *Opt. Commun.* **1978**, *26*, 41.

[55] H. ˇSípová and J. Homola, *Anal. Chim. Acta* **2013**, *773*, 9.

[56] A. Szabo, L. Stolz, and R. Granzow, *Curr. Opin. Struct. Biol.*, **1995**, *5*, 699.

[57] B. Liedberg, C. Nylander, and I. Lundström, *Biosens. Bioelectron.* **1995**, *10*, I–IX.

[58] A. W. Drake, M. L. Tang, G. A. Papalia, G. Landes, M. Haak-Frendscho, and S. L. Klakamp, *Anal. Biochem.* **2012**, *429*, 58.

[59] S. C. B. Gopinath, *Sens. Actuators, B* **2010**, *150*, 722.

[60] Biacore 4000, GE Healthcare Data File 28-9694-94 AB, Biacore, **2010**, pp. 7.

[61] A. Abbas, M. J. Linman, and Q. Cheng, *Biosens. Bioelectron.*, **2011**, *26*, 1815.

[62] J. Frischeisen, C. Mayr, N. A. Reinke, S. Nowy, and W. Brütting, *Opt. Express*, **2008**, *16*, 18426.

[63] Y. J. Jen and Y. H. Liao, *Appl. Phys. Lett.*, **2009**, *94*, 011105.

[64] J. Homola, *Anal. Bioanal. Chem.*, **2003**, *377*, 528.

[65] B. A. Prabowo, Y. F. Chang, Y. Y. Lee, L. C. Su, C. J. Yu, Y. H. Lin, C. Chou, N. F. Chiu, H. C. Lai, and K. C. Liu, *Sens. Actuators B* **2014**, *198*, 424.

[66] H. Y. Lin, J. H. Lee, M. K. Wei, C. L. Dai, C. F. Wu, Y. H. Ho, *et al.*, *Opt. Commun.* **2007**, *275*, 464.

[67] B. Y. Joo and D. H. Shin, *Displays* **2010**, *31*, 87.

[68] J. S. Lin, S. H. Lin, N. P. Chen, C. H. Ko, Z. S. Tsai, F. S. Juang, *et al.*, *Synth. Met.* **2010**, *160*, 1493.

[69] O. Hofmann, D. D. C. Bradley, J. C. deMello, and A. J. deMello, *Lab-on-a-Chip Devices with Organic Semiconductor Based Optical Detection in Springer Series in Organic Semiconductors in Sensor Applications*, Series: Springer Series in Materials Science, Vol. *107*, D. A. Bernards, R. M. Owens, G. G. Malliaras, (Eds.) **2008**.

[70] M. Pugliatti, E. Beghi, *et al.*, *Epilepsia* **2007**, *48*, 2224.

[71] N. Grossman, V. Poher, *et al.*, *J. Neural Eng.* **2010**, *7*, 13.

[72] A. Poghossian, S. Ingebrandt, *et al.*, *Sem.-Cell Dev. Biol.* **2009**, *20*, 41.

[73] P. Fromherz, *Progress in Convergence: Technologies for Human Wellbeing* (Wiley, **2006**), *1093*, 143.

[74] D. H. Kim, N. S. Lu, *et al.*, *Science* **2011**, *333*, 838.

[75] K. Svennersten, K. C. Larsson, *et al.*, *Biochimica Et Biophysica Acta-General Subjects* **2011**, *1810*, 276.

[76] V. Benfenati, S. Toffanin, S. Bonetti, G. Turatti, A. Pistone, M. Chiappalone, A. Sagnella, A. Stefani, G. Generali, G. Ruani, D. Saguatti, R. Zamboni, and M. Muccini, *Nat Mater.* **2013**, *12*, 672.

[77] Y. Nam and B. C. Wheeler, *Crit. Rev. Biomed. Eng.* **2011**, *39*, 45.

[78] M. Canepari, S. Willadt, *et al.*, *Biophys. J.* **2010**, *98*, 2032.

[79] F. Varela, J. P. Lachaux, *et al.*, *Nat. Rev. Neurosci.* **2001**, *2*, 229.

[80] J. Clark and G. Lanzani, *Nat. Photonics* **2010**, *4*, 438.

[81] R. Capelli, S. Toffanin, G. Generali, H. Husta, F. Facchetti, and M. Muccini, *Nat. Mater.* **2010**, *9*, 496.

[82] J. Y. Lee, J. H. Kwon, and H. K. Chung, *Org. El.* **2003**, *4*, 143.

[83] P. E. Burrows *et al.*, *Appl. Phys. Lett.* **1994**, *65*, 2922.

[84] P. van de Weijer and T. van Mol, *White Paper from EU FP7 Project – Fast2Light*, **2009**.

[85] T. van Mol, P. van de Weijer, and C. Tanase, *Proc. SPIE 6999* **2008**, *46*.

8

CONCLUSIONS

The potential of organic semiconductor-based devices for light generation is demonstrated by the commercialization of display technologies based on organic light-emitting diodes. Indeed, in terms of performance and reliability, organic-light emitting diode (OLED) technology is by far the most developed organic technology, and active-matrix OLED displays have been introduced into the market, primarily in the mobile consumer electronics sector. However, exciton–charge interactions and photon losses at the electrodes detrimentally affect OLED operation under high injection conditions. The close proximity of the metal contacts to the OLED light generation region induces losses due to absorption of the emitted photons. Moreover, the highly dense electron and hole currents converge in the light-emitting layer, where they form excitons and spatially coexist with them, leading to significant exciton–charge quenching. This mechanism is predicted to be greater than any other quenching effects at high current density and should be controlled to enhance OLED efficiency, brightness, and stability even further.

The organic light-emitting transistor (OLET) is an alternative optoelectronic device having the structure of a thin-film transistor (TFT) with the capability of light generation and efficient photon management. Thus, the OLET development is prompted by the possibility to exploit a transport geometry to suppress deleterious photon losses and exciton quenching mechanisms inherent in the OLED architecture and ultimately enable new display/light source technologies.

Organic Light-Emitting Transistors: Towards the Next Generation Display Technology, First Edition. Michele Muccini and Stefano Toffanin.

The light-emitting transistor configuration allows moving over a distance of microns the metal electrodes from the region where exciton formation and light emission occur, thus preventing exciton quenching and photon losses due to interaction with the metal electrodes.

The high mobility of carriers moving in a field-effect configuration also allows diluting the required charge-carrier densities in the device so to limit exciton–charge quenching and photon absorption by polarons. These key advantages of the planar structure coupled to the increased optical aperture of the electroluminescent pixels and to the possibility to easily incorporate photonic structures for light guiding, confinement, and extraction, position favorably the field-effect transistor approach to achieve efficient and bright electroluminescence from organic semiconducting materials. Based on these characteristics, OLETs are of immediate use in sensing platforms, optical communication, and integrated optoelectronic systems, which incorporate OLETs as a key active element.

Even more importantly, bright/multicolor OLETs have intrinsic structural advantages with respect to OLED for electroluminescent display fabrication. They offer an elegant solution to overcome the current limitations of the OLED display technology, which faces severe manufacturing hurdles and requires high-quality TFT backplanes to drive the OLED frontplane. The very nature of the OLET makes it a voltage-driven device and not a current-driven device as the OLED. This fundamental difference has huge implications when a display has to be designed, as it enables, in the case of OLETs, the use of lower quality TFT technology for backplane driving, for example, amorphous silicon or organic TFTs. The combination in a single device of the electrical switching and light emission functionalities reduces the number of switching TFTs to be realized in the driving circuit. In addition, OLETs offer easy-to-process device architecture that naturally avoids pin holes and shorts between injecting contacts, which is a major technological hurdle faced by OLEDs. All these aspects directly determine the production costs of the display and consistently point to OLETs as the next-generation display technology for both rigid and flexible applications.

INDEX

Organic Light-Emitting Transistors: Towards the Next Generation Display Technology,
First Edition. Michele Muccini and Stefano Toffanin.
© 2016 John Wiley & Sons, Inc. Published 2016 by John Wiley & Sons, Inc.